MARCELO CALDANA

SCIENCE
FORGOT
GOD

Translated by

Michael P. VanDyke

2020

The intellectual progress of humanity occurs by means of men who, with the lucidity of a meek and open soul, boldly expose their thinking to criticism.

Thomas Huxley (1825-1895), who was an English physician, writer, naturalist, and a solid opponent of Creationist Theory, with brilliant impartiality said, *"**Freedom of thought is the cause and at the same time the consequence of intellectual progress.**"*

Thus, reading this book will produce a seed that will make each reader thereof a contributor to the progress of human thought.

This book is dedicated to my teachers

of Science and of the Bible.

INDEX

A note from the author

I wrote this book for those interested in the fascinating subject of the origin of the universe and of life. I have used the most up-to-date scientific information and correlated it with similar biblical texts with worthy, impartial attention. In this way, we may conclude, after a long and thorough comparative study, that Science is unknowingly on the path of proving what the Bible teaches us to accept by faith.

I have studied for thirty five years with teachers of the Bible at my local church and with Masters of Science since my unforgettable days as an engineering student. I have always found disregard to be unfortunate, sometimes with faith in a creator god, sometimes with a scientist who consumes his days in search of answers to questions that are not only his own, but of all humanity. Science lives on men's doubts, trying to prove or disprove them. Faith does not require proof, but science thrives on it.

Chapter 1
Before the first day

The Earth is the greatest wonder of the solar system and probably of the Universe. It is perfect for human existence and it is a great place. It is unlike any other body in the solar system. There are absolutely diverse landscapes. It is special due to the temperature because it is at a perfect distance from the Sun. A few degrees more inclination of its axis would be enough to melt the polar ice caps and raise the temperatures to unbearable levels of heat, making human existence impossible. Only the Earth has water in its liquid state. The ratio of water to dry land provides the existence of millions of life forms. To this day, it is the only planet we know of where life really does exist. The Earth with life and man on it, is part of a vast cosmic world that goes far beyond what we can imagine.

Science believes that it arose from the process of forming the Sun. Photosynthesis made life become more complex. The influence of light on the process of creating life on Earth was decisive.

Let us now analyze a Bible passage and seek a technical correlation with what we have said above: *"In the beginning God created the heaven and the earth."*(Gen.1.1)This sentence is a statement and not a philosophical argument. The word *"beginning"* does not specify a time or an era. It does not temporally locate this creative act in the course of history. The scriptures do not present data for determining the time that led to the creation of *"the heaven and the earth"* – that which we call the universe. However, science categorically states that the Earth is 4.7 billion years old. This measurement has been performed by geochronologists using the radioactive dating method, based on the age of terrestrial rocks, using a radioactive clock. The hands of this clock are isotopes of the same element that contain different atomic masses, and the geological time is measured by the decay rate of one isotope in another. Among so many clocks there is Uranium-238 in Lead-206 and Uranium-235 in Lead-207. The earliest records of dating rocks by using this method are from the early twentieth century by Ernest Rutherford, the Nobel Prize winner in Chemistry in 1908, when he investigated the disintegration of elements and the chemistry of radioactive substances. Rutherford is considered the father of Nuclear Physics. If this book had been written in the nineteenth century, its author would have claimed, based on science, that the Earth was 25 to 75 million years old. It was a fabulously imprecise statement. However, as of 2020, science has become convinced that the earth is 4.5 to 4.7 billion years old. It is a much higher degree of relative precision, around 96%, although 200 million years are far beyond our imagination. The oldest fossils of life have been found in Australia and South Africa. These relics of blue algae are about 3.5 billion years old. It's a literally incomprehensible number, practically an eternity for those who live on average 80 years. If we take the Bible passages below and put them together with the scientific proposal of the age of the Earth, without further difficulties we will conclude that there has already been an eternity and that it will continue after man. *"..The everlasting Father"* (Isaiah 9.6) – *"For a thousand years in thy sight* are but *as yesterday when it is past, and* as *a watch in the night.* " (Psalms 90.4) - " *But, beloved, be not ignorant of this one thing, that one day* is *with the Lord as a thousand years, and a thousand years as one day"* (2 Peter 3.8).– *" But the mercy of the LORD* is *from everlasting to everlasting .."* (Psalms 103.17) (Past and future) – *"... before the world was"* (John 17.5)(past) - *"...whose goings forth have been from of old, from everlasting"*(Micah 52)(past) – *"Before the mountains were brought forth, or ever thou hadst formed the earth and the world, even from everlasting to everlasting, thou art God."– "Art thou not from everlasting, O LORD my God, mine Holy One?" (Past) – ".. thou art from everlasting."(Past) (Psalms 93.2)*

Without trying to understand what Eternity is or how it is possible to be eternal, and by adding the aforementioned texts to our reasoning, we can agree with science and its chronology without major conflicts. Stating that there was an eternity before Man and that there will be another after Man is not a heresy from the scientific point of view, nor from the Biblical one. It is as though science were already measuring eternity!

What is time?

In the Theory of Relativity, each person has their own measure of time. If time varies with the rotation of the earth, then time is stopped at the poles, where the meridians meet. If the day comes that we travel at the speed of light (approximately 300,000 km per second), for a traveler time will be 10 times slower than time for those who stay here on Earth. Every 365 days and 6 hours our clock shows that everything will repeat itself – another winter, another summer and so on. But this is not time. We age faster or slower depending on the direction of rotation and the speed with which we travel around the Earth. If there is any physical relationship between the rotation of the Earth and time, the Earth will stop at the beginning of Eternity. Then there will be only one weather station at each end of it. This reasoning seems illogical, but it will not be if we accept that time does not depend on the rotation of the Earth. Albert Einstein stated that time had a beginning. It is a fabulous statement that is naturally not limited to our planet and human existence. But if we exercise our imagination a bit by taking the Bible passage about the emergence of the *"..two great lights; the greater light to rule the day, and the lesser light to rule the night"* (Gen. 1.16) (the Sun and the Moon) and accepting that time on Earth is marked by the movements of these celestial bodies in relation to it, we will arrive at another conclusion: that time, at least for mankind, actually began on the fourth day of creation, thus endorsing Einstein's thesis by stating that "time had a beginning".

Is eternity an invention of the Bible? Amidst the stars there are places called "black holes". There, time does not exist and light does not come out from within them. Although General Relativity cannot yet define an equation for the end of time, in these places Eternity does exist. But let us return to the initial text *"In the beginning God created the heaven and the earth."*(Gen.1.1) In this sentence of the scriptures I emphasize the words "the heavens and the earth". It is not written as "the heavens and the moon or the heavens and Mars". Note also that the expression "the heavens" is not in the singular. It is evident and at the same time odd that the text puts the Earth in the context of the universe and consequently of all creationist theory. As for "the heavens" I have learned in Bible schools that saying heaven or heavens has the same meaning. However, upon seeking a possible correlation in science, the understanding seems different. Here's what the English physicist and astronomer Martin Rees says in his Multiverse thesis: *"Our universe is only one in an incalculable series of universes existing in different dimensions of space-time." "What I'd like to know is whether these universes are based on physics and turn out to be correct, and whether the different universes would be governed by different physical laws."* We can also cite the String Theory about how all the different stable vacuums of the string panorama emerged at various points in the universe, forming innumerable sub-universes. Moses, attributed to the authorship of the book of Genesis by Jews and Christians, wrote: *"Behold, the heaven and the heaven of heavens is the LORD's thy God"* In the book of Job, the first of the Wisdom books of the Old Testament canon , whose author is unknown, and which was written at an uncertain date, there is the following phrase attributed to the creator of the universe itself." .. *whatsoever is under the whole heaven is mine."*(Job 41.11)

Einstein even wondered whether there was any choice in the cosmic blueprint: *"What really interests me is whether God could have created the world any differently; in other words, whether the requirement of logical simplicity admits a margin of freedom."*

In other words, if creation was random, the result of one natural or evolutionary accident after the other, or if it were with well-defined laws, would the creator have known what would happen in one million, in ten million, or in a hundred million years. !?

In the Old Testament the prophet Jeremiah, whose life and work are historically known in depth, in the book he wrote in the 7th century BC, wrote: "Thus saith the LORD, which giveth the sun for a light by day, *and* the ordinances of the moon and of the stars..." (Jeremias 31.35)These "ordinances" (fixed laws) may refer to those, which Kepler enunciated about the movements of celestial bodies, or to be much more comprehensive considering the determinant influence of these movements on the planet Earth.

Chapter 2
The Primordial Earth

Earth is a mutant planet, where mountains literally move, huge lakes turn into deserts, and large stretches of land that were once dry and barren begin to bear fruit. The magnetic poles move, and powerful underground forces shape the surface, demonstrating that the planet Earth is geologically alive. This feature, which has already been well proven by science, is endorsed by a statement from the Bible affirmation which quite categorically tells us that "the earth was without form, and void." (Genesis 1:2) The verb "to be" is in the past, indicating that in that specific instant of the biblical narrative the Earth had a shape that is different from the present one. What shape does chewing gum have after it has been chewed? It is not a cylinder, a cone, or a cube or any other defined geometric shape. One could say that it has no shape whatsoever, for there is no way to describe it. There is a theory which states that the shape of our planet is the result of a successive irregular "heaping up" of small and large celestial bodies that were attracted by the gravity of the earth, and this is called accretion. The clause "was without... void" calls for special attention to each one of its words. The Earth was part of a setting where only it was empty. There is no indication in the Bible of the composition of its atmosphere. "Void," or empty, does not mean the absolute absence of everything as though a vacuum surrounded the Earth. When one says "the bottle is empty" it is a mistake because the inside of it is filled with air, which is a gaseous fluid and occupies a location in space. In order to fill it with water, for example, the air must leave and relinquish its space. Therefore, by reasoning with the same logic, saying that the Earth was void must be followed by asking "void of what?"

The answer could be "void of anything visible". When we relate this expression to the other stages of creation, according to the chronology of the biblical narrative, we will deduce that the Earth was initially void of light. "And God said, Let there be light ..." (Genesis 1:3). According to this famous clause, at that moment in the history of the universe, and of the Earth in particular, "light" had not existed and by these words it was called upon to exist, that is, it came into existence.

But what is light? Why was it created? And why did that occur on the first day? According to a dictionary of the English language, light is an electromagnetic radiation of any wavelength (4000 – 7800 A) that travels in a vacuum with a speed of 299,792 kilometers per second (about 186,282 miles per second). I must add that light is a form of energy called Electromagnetic Energy. It is composed of photons, or quantum of light, which are small particles of energy. In 1925 Albert Einstein arrived in Rio de Janeiro and lectured at the Brazilian Academy of Sciences on the nature of light, comparing wave theory and quantum theory. The Danish physicist Niels Bohr, the Scottish physicist James Clerk Maxwell, Hendrik Anthony Kramer, John Slater, the Germans Walter Bothe, Hans Geiger and their compatriot Werner Heisenberg (1901-1976) researched the same subject for 50 years, and they disagreed on some points on this issue of light and its behavior as a wave and as a particle.

If Genesis 1.3 "let there be light" is interpreted as letting there be brightness and luminosity as of that historical moment called the first day, both science and religion will be faced with yet another great question. The light that illuminates the Earth comes from the Sun. Without it mankind would be in darkness and life would be unfeasible on the planet. What light could have been meant by, as the Sun was not created until the fourth day? "And God made two great lights; the greater light torule the day, and the lesser light to rule the night . And God set them in the firmament of the heaven to give light upon the earth."(Genesis 1:16-17) The Interstellar Boundary Explorer (IBEX) was launched by NASA on October 19, 2008. Science objective is to discover the nature of the interactions between the solar wind and the interstellar medium at the edge of our solar system. Dr. McComas, IBEX principal investigator and assistant vice president of the Space and engineering Division at Southwest Research Institute, coordinator of this mission, expressed himself about some information sent from the spacecraft:"Over the past six years, this fundamental work focused on our place in the solar system has become the gold standard for understanding our sun, our heliosphere and the interstellar environment around us," said David McComas, principal investigator of the IBEX mission at the Southwest Research Institute, or SwRI, in San Antonio, Texas. Instead of a circular and homogeneous heliosphere, the data indicate the presence of a narrow, extremely 'bright band winding through the sky - where bright refers not to any light, but to a high concentration of energetic neutral atoms(ENAs)... We had expected to see small and gradual spatial variations at the interstellar boundary, some ten billion miles away. However, IBEX is showing us a very narrow band that is two to three times brighter than anything else in the sky .... The results of IBEX are absolutely extraordinary, with emissions that cannot be explained by any theory", concludes Dr. McComas.

"There is no light", someone screams when the lights go out. In fact, there is a lack of electricity to be transformed into light by a light bulb. So why was light created on the first day? Now the answer is getting easier. Nothing, absolutely nothing, works without energy and according to the first Law of Thermodynamics it is only conserved, changing its form – chemical, electrical, wind, thermal, mechanical, kinetic, potential and luminous energy. Gasoline contains chemical energy and an engine turns it into mechanical energy for wheels, electrical energy is used for illumination and the waste becomes thermal energy, that is, heat that returns to nature. The earth and the universe do not live without energy. Everything is in motion; nothing is static. The earth moves in its orbit at 67,000 miles per hour, the Sun moves at 450,000 miles per hour and everything is based on energy that is always transforming and changing, without ceasing to exist and without losing its intensity.

Therefore, if it is difficult to believe without proof, at least there is technical coherence in creation. The subsequent creative acts described in the Bible could not have occurred without energy being created on this first day. Could the "Big Bang" be the exact moment of "Let there be light"? Isaac Asimov, one of the most brilliant minds to have ever passed through this planet expressed himself about energy in the following way.

"In a way, of course, we could argue that the energy of the Universe (including matter as a form of energy) has always existed and always will exist, since as far as we know, it is as impossible to create energy out of nothing as it is to destroy it to nothing. This implies, we may conclude, that the substance of the universe – and therefore the universe itself – is eternal."

Chapter 3

The Magic Wand

Cinema is a fantasy world, a world in which everything is possible. It creates super-heroes capable of turning back time by reversing the earth's direction of rotation, creating a powerful green man, mutant villains, aliens of all shapes and colors, immense spaceships that silently hover over a metropolis, magical potions, and webs coming out of wrists, and all this is done without forgetting "once upon a time" and "they lived happily ever after". Film producers are able to bring into existence a world without the laws of physics, a world above the forces of Nature.

With careful impartiality, I have read selected texts from books, journals and specialized magazines that may somehow be related to the creation of the universe and of man. I have reluctantly come to the conclusion that the common understanding of creationist theory points to an instant creation by means of a few orders. If we take the celebrated text from Genesis "Let there be light: and there was light" (Genesis 1:3), one's first impression is that of instantaneousness. However, the Bible is a book of contexts and its verses must not be interpreted in an isolated manner, under the risk of not reaching an understanding of what the author wanted to pass on to the reader. Further on in Genesis, another phrase that serves to illustrate this reasoning is "Let there be lights in the firmament of the heaven, ..". Notice that the verb "to be" is in the same tense upon expressing the order of existence of the "lights", referring to the sun, moon and stars. It is now known that a star takes approximately two million years, forming from gases and stellar dust under vigorous variations of temperature and pressure, before it appears in the sky the way we see it here from Earth. Another example that characterizes this absence of instantaneity is in the other verses of the same book, which continue to describe creation. Among them is this one, "Let the earth bring forth grass, and herbs yielding seed, [and] the fruit tree yielding fruit..." (Genesis 1:24). The order was given for the earth to produce grass, herbs, and fruit trees with the natural resources of its soil and of its atmosphere at that time. At the moment of the aforementioned biblical narrative the sun did not yet exist, which would make germination technically impossible. But there was something latent so that in due time its produce would spring forth from the ground. Some law which is unknown until today was acting silently within the planet and externally in its primordial atmosphere. There are no indicators of the duration of each of these events, because the word "day" in the context of the Bible has other applications such as a period of history or an age, as well as the 24-hour day itself.

I understand the words "... image and likeness..." as a similarity of order and principles, logic and planning, for only then could a plan be carried out on these pillars. Even though I am an engineer who questions the reasons for things each and every day because it is my official duty, I cannot see a casual universe, much less a god that is a creator with a magic wand in his hands. Although my faith resists my logical reasoning, I do not live with this conflict, for I am relaxed as I dwell on the organization of the universe and on its symmetry, which is its unyielding and unvarying equilibrium. An accident of Nature that is a casual, extraordinary fact which is not part of the everyday life of the planet generates disorder like a tsunami or a volcanic eruption, although these events are a natural consequence of the powerful forces that act externally and internally on our planet in the pursuit of balance.

There is not one single book, one single scientist or one single theory that offers arguments with sufficient and concrete basis regarding the creation of the universe and man. Miraculous, delirious theories without the least scientific consistency are irresponsibly effused into a society that sits mutedly fascinated with magazines and television sets. Finding out how nature works is a task for science. However, the explanation of the creation of the world seems to have become a passionate race toward proving that God the creator does not exist. In my judgment, that is partiality, which is not science.

Chapter 4
Faith in Science

At no time did scientists have as many images and data of the universe at their disposal as they had at the beginning of the twenty-first century. Physicists, astrophysicists, astronomers, cosmologists, astrobiologists and so many others in several fields of scientific research have their laboratories crammed with information from deep space, which has been collected by probes and powerful telescopes that scour the universe at their fingertips. However, every piece of incoming information brings about new questions, but the answers do not come along with them. At one time science was delighted with the scarce data that did come in from outer space. Today science is as though it were numbed by an avalanche of information without any answers. It is at the foot of a mountain being buried by questions that come sovereignly downhill, disregarding diplomas and curricula. Society in turn waits meekly for some answer, accepting with a childish stance whatever reaches its ears, because it has faith in science. It has faith in men who speak about the billions of years of the universe with the same naturalness as those who speak of their recent childhood. It has faith in men who talk about distances of millions of light-years, at a time when an airplane takes twelve tedious hours to cross an ocean.

The secret of science is observation and logic is its compass. Emotions or even discrete trends can invisibly mask the interpretation of a scientific observation. Atheism is a religious trend in the same way that genuine faith is. Faced with such a categorical declaration as "Let there be light" a believer exclaims, "how powerful God is!" An unbeliever says, "it is a fantasy." A scientist, however, whether he is a believer or an unbeliever, must ask, "how could this be technically possible?" He neither doubts it nor believes it until he proves one statement or the other to be true. His logic, which both the atheist and the believer lack, should lead him to neutrality. In principle he doubts nothing, but questions everything. Success in his research depends on his ability to observe nor on his faith or his atheism. Science is proof and impartiality. Johannes Kepler was reportedly a Lutheran. Nicolaus Copernicus was a graduate in Canon Law at the University of Bologna in Italy.Charles Darwin studied Theology at Cambridge. Isaac Newton was an Anglican Christian. Albert Einstein mentioned a god throughout his life and career, questioning his intervention in the universe and the life of men. It is said that Pasteur was a Catholic, but that he did not go to Mass. James Watson said he was lucky to have been raised by a father with no religious beliefs and a nominally Catholic mother.

I have heard that the advances of science are leaving little room for God, with people questioning whether religion will survive. Why is there so much bias in qualifying religion as "clouds of ignorance and superstition", "mystical speculations" and "anthropocentric superstitions". The same questioning, investigative and impartial logic should be adopted for any theories, considering them to be credible until the moment of their actual verification, without any demerit to any of them. Nature works in accordance with scientific principles that are absolutely indifferent to what men think. Before Newton enunciated it, gravity was already acting upon believers and unbelievers. Before Kepler, celestial bodies were already moving meticulously in the sky, dazzling the men who did not know the perfect and inflexible laws that governed them and that will forever rule their repeated ways. Life and time will continue to be two illustrious strangers as long as the prejudice of religious people and scientists remains as though they were blind. Faith, by its biblical definition, is "the evidence of things not seen." (Hebrews 11:1) Nobody saw God creating life and the universe.

Nobody saw the Big Bang. Nobody saw the first molecule or the first form of life. Having faith in science is equal to having faith in scientists, what they study, what they conclude, and what they disclose as the truth, even if in the next decade new discoveries radically alter what was expressed as truth. Faith is believing in the words of men that are written in books – in a religion book or a science book, in a book that chronicles creation, or in an argument for evolution or the Big Bang. The book that narrates creation describes a principle; the books that argue with coherence and intelligence in favor of evolutionary theory say that evolution needs a principle in order to have evolved from it. They present no argument about what the principle could have been like.

Whether one accepts it or not, it is undeniable that science requires faith which must be greater than "... a grain of mustard seed."

Chapter 5
Anthropocentrism

Nicholas Copernicus dedicated four decades of his life to the study of astronomy, and he died in 1543. His conclusion was that the Earth is not the center of the universe and that it has no special role in Creation, quite unlike the dominant thought had been for the previous four millennia. His studies were truly revolutionary, since Christian Theology had preached for 1,500 years that the Earth was the center of the universe, the center of Creation – that the sun, the moon, planets, stars and other celestial bodies revolved around the Earth. Aristotle preached that the immovable Earth was at the center of the universe However, one of the most brilliant scientists in the history of mankind, the German astronomer Johannes Kepler discovered mathematical laws of celestial orbits that are applicable to any stellar system, proving in 1609 that Earth was not the center of the universe. With the extremely limited technical resources available in their respective ages, being correct or incorrect in their conclusions, these brilliant minds have demonstrated a fabulous ability to observe and conclude, coupled with extraordinary courage to present their conclusions, which most often were in conflict with religious power and with the scientific community of their respective era.

In our present days there is not a more comfortable relationship between science and religion. Nonetheless, the incredible equipment available to researchers, such as the monumental LHC (Large Hadron Collider) which is able to collide and study particles weighing one trillionth of a gram where 0.000000000001 second is an eternity and the cinematographic Hubble radio telescope, which penetrates into deep space where man will never arrive personally at the frightening distances of 12 billion multiplied by 10 trillion kilometers, provide a fabulous amount of information that feeds scientists daily in their labs, the scenario still looks the same. Religion, in turn, continues to believe and preach that there is a divine creator of the universe. Attentive, quiet and always convinced, it accompanies the extraordinary technological development that humanity has enjoyed over the last 20 years in medicine, astronomy, communication and, in a general way, the quality of life of human beings. Science, on the other hand, delights in providing humankind with a life that is admittedly more comfortable, safer, and more effective. Nevertheless, the fundamental questions remain unanswered while nothing has changed and nothing has been added other than new, fantastic theories in the phase of dedicated studies.

Expressions of feelings such as tears, smiles, love and hate, for example, continue to be accepted as the result of an evolutionary accident that emerged from nowhere, was initially transformed into inanimate matter and then, mysteriously, into life that spread throughout the planet Earth.

Curiously, the first sentence of the Bible is "In the beginning God created the heaven and the earth." It is noteworthy that it is not written as "... the heaven and Mars" or "...the heaven and the Moon", making it quite clear that the planet Earth stands out in the context of the Universe. The searches for extraterrestrial intelligent life or even for any sign of life have remained silent, with no answers and no clues for fifty years. The Brazilian writer Marcelo Gleiser, who is a professor of Natural Philosophy and of Physics and Astronomy at Dartmouth College and a researcher in theoretical physics, has called the situation of man in the universe "cosmic solitude". This solitude makes man unique as well as the planet Earth itself. The other known celestial bodies have extreme temperatures that are icy or very hot, without providing the slightest conditions for the existence of life that is at least similar to life on Earth.The earth is a garden. Man is the greatest spectacle of life in the universe – an accidental spectacle from the viewpoint of science and a spectacle that was meticulously designed from the viewpoint of the Bible.. Regardless of the position of the earth in relation to the other celestial bodies, for this reason Man is the center of the Universe, whether seen as a complete initial creation or as a wonderfully evolved creative accident.

Chapter 6

The Second Day

"And God said, Let there be a firmament in the midst of the waters, and let it divide the waters from the waters. And God made the firmament, and divided the waters which *were* under the firmament from the waters which *were* above the firmament: and it was so. Gen1.6-7 In this passage of the sacred scriptures I want to emphasize the words <u>under</u> and <u>above the firmament</u>. According to dictionaries of the English language, the word firmament in Astronomy means a visible celestial hemisphere, sky or celestial arch. In another sense, it can mean base, support, or foundation. The book of Genesis was written, probably by Moses, around 1450 BC. He is not considered the intellectual author. Moses lived his youth in society and in the Egyptian court. The Bible states that he was educated in "all the science of the Egyptians. By science, in this case, we mean reading, writing, classical literature, archery and physical activities. We do not have knowledge of whether Moses had a vision and then reported it in his own language, as his great understanding of culture allowed him to interpret what he saw or whether it was a literal transcription from the creator himself. It is less likely that he found some earlier historical records and compiled them. Regardless of Moses's academic formation or whether he actually was the author of the book of Genesis, the account is shown to be coherent and technical. If we understand that firmament means base, support or foundation, we can easily conclude that the water cycle of the Earth was being defined and physically formulated at that time. The first law of Thermodynamics, the law of conservation of energy, would have already been in operation, thus heat would have already been causing wind when encountering masses of air at lower temperatures. Clouds were already capable of supporting the sky, the firmament in the sense of astronomy. The waters would come to evaporate, turn into clouds and return to earth in the form of rain, snow or ice. In the words of the book called Ecclesiastes 1.7 is the summary of this cycle:

" All the rivers run into the sea; yet the sea is not full; unto the place from whence the rivers come, thither they return again."

Chapter 7
The Third Day

The word "terra" came from the Latin word "ters" into the Portuguese language in the thirteenth century to designate ground, soil, and dry. For the word "earth" in the English language and for the word "erde" in German, the root comes from the Greek word "Éraze" , and it means "on the ground". Regarding its shape, 2500 years ago Pythagoras stated that the Earth was round, although he did not have the scientific means to confirm that. He based his theory on the image of a ship that disappears on the horizon, starting at the hull and continuing to the tip of the mast. Interestingly enough, 300 years before Pythagoras the prophet Isaiah was referring to the creator when he wrote in chapter 40 of his book, "[It is] he that sitteth upon the circle of the earth..." (Isaiah 40:22) In a macro view, the earth is already well known as a planet in terms of sea currents, winds, the soil, the climate, and its physical relationship with the part of the universe that surrounds it. However, with the degree of accuracy we enjoy today, this knowledge is still a very recent phenomenon. It is an extremely modest sort of knowledge in its predictability, limiting itself to knowing its predominant movements, albeit without the slightest capacity of reaction. This is due to the fabulous, immeasurable and magnificent forces that make our planet work. The waves of the sea do not cease, the winds only change in direction and intensity, whereas the air of a metropolis changes temperature in a matter of a few minutes. If we compare the power of the equipment we use to control the temperature in our office, for example, and if we dare design an air conditioning unit for our entire city, we will be faced with numbers that would not fit in our calculators and with so much energy that we would not have it by uniting all the power plants on the planet. The numbers would be terrifying. The energy needed to move a fabulous tectonic plate that causes earthquakes and tsunamis is literally incalculable. The Saturn V rocket, which propelled the Apollo capsule to the moon with three astronauts aboard, required approximately six million horsepower just to get off the ground and reach the speed necessary to begin that historic voyage. Michael Gurnis, a geophysicist at the California Institute of Technology, in his article entitled *Processes that Carve the Earth*, wrote that "Enigmatic cases of rising and sinking have occurred several times in the past on portions of the earth's surface, the size of continents. The southern part of Africa has been raised about 300 meters in the last 20 million years, for example, and the highest peaks of a sunken continent today form the islands of Indonesia. The scientific community is already unanimous in accepting the concept of so-called primordial oceans, the oceans in the primitive Earth. In 1936 the Russian biochemist Aleksandr Ivanovitch Oparin wrote in his book *The Origin of Life* that "the primordial oceans functioned as a huge chemical laboratory." This correlation of science with biblical phrases becomes fascinating as we add new texts with impartiality and criteria. For example,"For he hath founded it upon the seas, and established it upon the floods." (Psalm 24.2), referring to the earth and by "floods" it can be deduced that it means sea currents. This biblical statement in and of itself does not reveal that there was an entire ocean covering the whole earth, but let us add this other statement, "And God said, Let the waters under the heaven be gathered together unto one place, and let the dry land appear: and it was so." (Gen.1.9) It may be deduced that the primordial oceans covered the whole earth, and that at a certain moment in history the sea level changed, causing the "dry portion" to appear. See what was said about this by Dr.Roger L. Larson, a geologist from Iowa

University, with a Ph.D. in Oceanography from the Scripps Institution of Oceanography of California, an associate researcher at Lamont-Doherty Earth Observatory, Columbia University, and Professor of Oceanography at the University of Rhode Island, with 15 expeditions on scientific ships.

". The Middle Cretaceous period (between 120 and 125 million years ago) was characterized by several deep anomalies resulting from the superswell. The first and probably the least controversial [point] is the global rise in sea level of 250 meters or more compared to what it is today. Considering that the total amount of marine water on the planet is constant, the rising ocean level is simply a reflection of a corresponding rise in the level of the ocean floor. In the Middle Cretaceous, many currently dry areas were underwater; For example, my hometown in Iowa was at the bottom of the ocean. When the water receded, it left deposits of limestone and chalk, including the famous White Cliffs of Dover in England." In Lisbon, Portugal, there is an Oceanarium that reproduces the idea and sensation of one single ocean. The commercial for this tourist attractions expresses it like this: "Travel to the Primordial Ocean." The arrangement of the different aquariums in the building aims to convey the message that there is only one global ocean. For several reasons, including the current position of the continents, the ocean has been artificially divided by man into several oceans and seas. In reality the marine environment is a continuous environment, in which the only existing barriers are physiological ones that impede the free movement of animals around the four corners of the planet.

"Let the waters under the heaven be gathered together in one place, and the dry portion appear. The dry portion God called earth and the gathering of waters, called he

Seas."(Gen.1.10)The emphasis on the waters under the heavens being in one place may be due to Pangea.

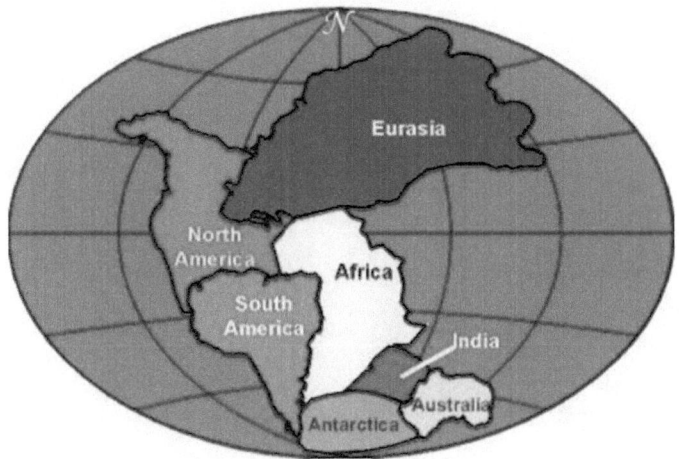

"**Pangea** was surrounded by a single ocean called Pantalassa.[1] The hypothesis was initially suggested in the early twentieth century by the German meteorologist **Alfred Wegener**, creating a great controversy among the scientific field at that time. As the starting point for his theory Wegener pointed out that the contours of the coast of South America were similar to those of Africa, which would form an almost perfect fit. However, it was not this fact that was used as the scientific basis of his theory, but rather the comparison of the fossils found in the Brazilian and African regions. As these animals would not have been able to cross the ocean at the time, it was then concluded that they would have lived in the same environments in ancient times. This was not confirmed until 1960, 30 years after Wegener's death." Pangea could explain how there was a single continent in one place. It can be deduced that it must have been surrounded by a single ocean. But at the end of the sentence is the word "seas", which is in the plural form. It is highlighted that the dry portion was all in one place, but as for the seas nothing is emphasized and there is no mention that there was only one. The so-called Pantalassa could explain these "seas".

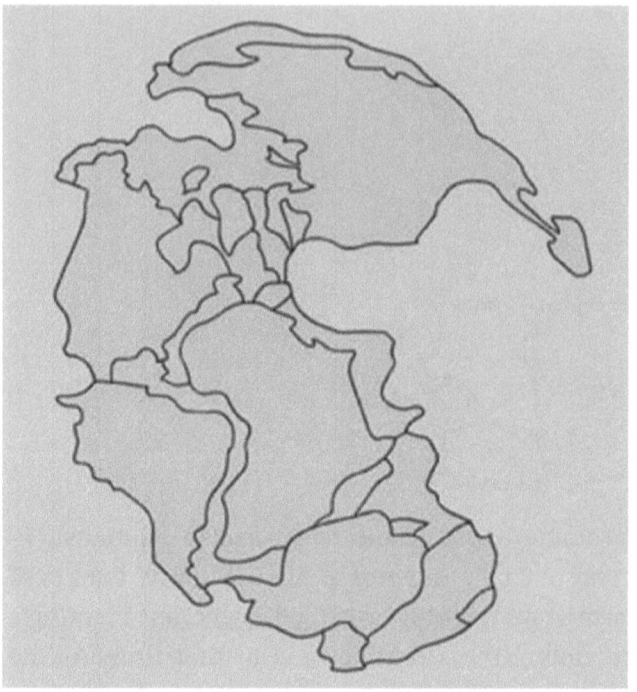

   "Pantalassa, (from the Greek 'pan + talasso', which means 'all the seas') also known as Panthalassa or Pantalassic Ocean, was the once vast global <u>ocean</u> that surrounded the <u>supercontinent Pangea</u> during the <u>Paleozoic</u> and early <u>Mesozoic</u> <u>eras</u>. <u>It included the primitive Pacific Ocean, to the north and</u> <u>west, and the Sea of Tethys to the southeast.</u> It became the present Pacific Ocean after the closure of the Tethys Sea basin and the fragmentation of Pangea, which led to the creation of the <u>Atlantic, Arctic and Indian</u> Ocean basins. Panthalassa is sometimes called the Paleo-Pacific ('old Pacific'), due to the fact that this ocean evolved from it." Therefore, alongside the "dry portion" there were "Seas", as the author of Genesis literally narrated. Nonetheless, the biblical narrative of the third day did not end there. A new order emerged in the following manner

"And God said, Let the earth bring forth grass, the herb yielding seed, and the fruit tree yielding fruit after his kind, whose seed is in itself, upon the earth:" (Gen. 1.11)In order to make a cake, a recipe and the components thereof are needed. When all the ingredients are mixed in a certain condition, putting them in the stove and waiting for the oven to produce a tasty cake will suffice, even though what was put there had neither the appearance nor the aroma and the flavor of a cake. The imperative sentence "<u>Let the earth bring forth</u>..." is of a fabulous scientific magnitude because it implies the understanding that the technical scenario for spontaneous generation of some vegetables was already present on the earth at that particular moment in its history. The order to bring them forth was given to the earth itself. The so-called Spontaneous Generation of Plant Life mustn't be confused with the Spontaneous Generation of Living Creatures and Man.. The Theory of Abiogenesis came up as an attempt to explain the beginning of life on earth, considering as life all living beings, whether they are unicellular or man himself. If it were proven, it would endorse Evolutionary Theory, but Louis Pasteur carried out several experiments and concluded otherwise. It was in the second half of the nineteenth century that abiogenesis suffered its final blow. Louis Pasteur (1822-1895), a French scientist, prepared some meat broth, which is an excellent culture medium for microbes, and subjected it to a careful sterilization technique with heating and cooling. This technique is known today as "pasteurization". Once sterilized, the broth was stored inside a "swan-necked" flask. Due to the long neck of the glass flask, the air entered the flask but the impurities were retained in the curve of the neck. No microorganism could get in to the broth. Thus, despite being in contact with air, the broth remained sterile, proving the absence of spontaneous generation. Several months later, Pasteur exhibited his material at the Academy of Sciences in

Paris in 1864. The broth was perfectly sterile. The theory of spontaneous generation was completely discredited. "For as the rain cometh down, and the snow from heaven, and returneth not thither, but watereth the earth, and maketh it bring forth and bud, that it may give seed to the sower, and bread to the eater..." (Isaiah 55:10) The way this is reported in the Bible is technically consistent with the chronological order of creation.. Energy was created on the first day, water was created on the second day, and on the third day earth was created in order to germinate seeds. But there was still something missing for underground germination to thrive, become alive and grow — the Sun. Nothing on this planet flourishes and remains without the light of the Sun The natural consequence — fruit — was inevitable. For while the earth was "producing" germinated seeds in its subterranean layers, the Sun, which is an indispensable part of life, at least of life we know it, was about to appear.

Chapter 8
The Fourth Day

"**L**et there be lights in the firmament of the heaven to divide the day from the night; and let them be for signs, and for seasons, and for days, and years...the greater light to rule the day, and the lesser light to rule the night..." (Genesis 1:14) Here the biblical quotation is "<u>Let there be lights</u>", whereas in the previous one it was "<u>Let there be light</u>". The word "light" has the following meanings, "Anything that emits light, that shines, a beacon or a star." There is no specific mention of the Sun and the Moon in this narrative. However, another biblical passage is quite clear. "To him that made great lights... the sun to rule by day... the moon and stars to rule by night..." What is the Sun? The American astronomer Donald Howard Menzel estimates that 81.76% of the volume of the Sun is hydrogen, 18,17%, leaving 0.07% for the rest of the elements. The Sun loses 4,600,000 tons of mass every second. Also, every second 630,000,000 tons of hydrogen are burned.

Its mass is estimated at 2,200,000,000,000,000,000,000,000,000 tons, that is, 333,500 times the mass of the earth. At 150,000,000 km away from the Earth, the Sun is a second generation star, the result of an earlier activity.

The English astronomer Fred Hoyle and the German astronomer Carl Friedrich von Weizsacker, among others, agree in postulating that the solar system, both the sun and also the planets, were formed by a single process. In other words, if the earth is 4.7 billion years old, the entire solar system (including the sun) as it now exists, is also 4.7 billion years old. According to the Bible, all these celestial bodies came to exist on the same day: the fourth one. Therefore, both Science and the Bible recognize that all celestial bodies, at least those of the solar system, appeared on the same "day" or in the same era.

In order to understand the formation of the stars, let us now go back to the second day. "And God called the firmament Heaven". (Genesis 1:8) Stars are not easily visible while they are being formed. When a star appears it is not ready in the same way that one sees it in the sky. Initially it is a conglomerate of cold gases and dust that goes through stages of contraction and heating, which radiate light from the infrared (invisible) band to the ultraviolet range. A star which has the mass that our sun has would perhaps take 2 million years to reach the point that we see it today. A star is born, it develops, and it dies. *"In general, astronomers agree that stars in the early stage are vast clouds of (interstellar) gas and (intersidereal) dust. Millions of years ago, when the galaxy was being formed, this stellar raw material must have been plentiful."* *"The observation we all make that the sky is dark at night is very important. This means that the universe may not have always existed in the state in which we currently see it."* Something must have happened in the past to make the stars light up a finite time ago." Can one then deduce that the stars began their process of appearing on the second day – in the formation of the "firmament", but not reaching its splendor as we know it today until the fourth day, which means two days that are perhaps one million years long apiece!? "Undoubtedly, the constitution of the vast majority of stars is remarkably uniform. Like the Sun, most stellar material is made up largely of hydrogen and helium." Curiously, the biblical narrative gives the Sun prominence among millions of other stars in the Universe that are brighter and often much larger. What could the reason be for its prominence? "Let there be lights in the firmament of the heaven to divide the day from the night; and let them be for signs, and for seasons, and for days, and years...the greater light to rule the day, and the lesser light to rule the night..." The sun became the reference for a day, for years, and for the seasons of the year. As we know, one year is a complete revolution of the earth around the Sun and one day is a complete revolution of the earth around its axis. In other words, time came to exist due to the existence of the sun, at least in the solar system – for us here on earth. Still regarding the appearance of time, let us read what science has to say. "... All the matter of our cone of light from the past is bound in a region whose boundary shrinks down to zero. Therefore it is not surprising that I (Stephen Hawking) and Penrose (Roger Penrose is an English physicist and mathematician, an emeritus professor of mathematics at the University of Oxford) have been able prove that in a mathematical model of general relativity time must have a beginning... Our article, which proves that time had a beginning, won the second-place award of the Gravity Research Foundation in 1968 (Hawking and Penrose, 2010)

Chapter 9
The Fifth Day

"**A**nd God said, Let the waters bring forth abundantly the moving creature that hath life, and fowl that may fly above the earth in the open firmament of heaven." (Genesis 1:20) This clause, "Let the waters bring forth (to populate, to stock, to fill, to establish oneself, to become populated)..." was general for at that moment in time, the oceans and the bodies of fresh water already existed on earth. The order to "Let the waters bring forth..." was discreet yet expressive, indicating a spontaneous, impersonal act. So this biblical phrase does not ascribe to God a creative act of the kind came into existence instantaneously, but submits the entire creative process to Him, His watchword, His agency, and His sovereignty over the earth and the universe.

"And God created great whales, and every living creature that moveth, which the waters brought forth abundantly, after their kind, and every winged fowl after his kind ..." (Gen. 1.21)

Here the verb used was *create*, with the following meanings in the English language: bring into existence, cause, design, lead to, occasion, produce, generate, form

According to the Holy Scripture, it is at this point, on the fifth day, that life begins in all its kingdoms, and along with it, all its fabulous charms and impenetrable mysteries. For scientists, life is a hindrance in their laboratories; it is a ghost that haunts them every day. Every smile that is received and every tear that is shed violate the intelligence of one who imagines a scientifically formulated life. In the human body, a smile, which expresses joy, is what moves the largest number of muscles simultaneously.

The muscles can be explained, and one can perhaps explain the brain, whose neurons signal movement to the respective muscles involved, but the brain does not explain the joy that brings about a smile nor the pain that causes one to cry. There is no theory that formulates an act of heroism. There is no laboratory that can find a formula to be applied to men that makes them love or hate. At the time of death the muscles, brain and blood are still in the body, which loses life in a fraction of a second. For science, the heart has stopped beating.

So then, what is life? In 322 BC, Aristotle said that life and nature were one and the same, acting under the guidance of God. In the seventeenth and eighteenth centuries, the focus was on nature and how it works. At the end of the eighteenth century came the movement called Vitalism, which differentiates between living beings and inanimate objects. In the nineteenth century, along came biology and with it the question: what is life? The German philosopher Immanuel Kant thus defined it, "— life is being born, growing, living and dying." But stars also are born, grow and die. Therefore, this definition of life was wrong. In 1976, the new definition that emerged through Jacques Monod (The *Nobel Prize* in Physiology or Medicine 1965 was awarded jointly to François Jacob, André Lwoff and *Jacques Monod*) "life is a system capable of perpetuating itself over time by keeping its own information." That is, the power to reproduce and transmit unchanged information about its own structure – this definition was called Invariance. Knowledge of life is reduced to knowledge of the structure and functioning of genes. Proteins maintain life. Nucleic acids (DNA and RNA) maintain the inheritance of information. Life is the harmony and simultaneity of proteins and nucleic acids. But which of the two came first? The question is the same as the classic "which came first, the chicken or the egg?" Robert Williams of the University of Oxford in England and João José Fraústo da Silva of the Technical University of Lisbon say, "Those who do state-of-the-art research, whether in microbiology or in physics, do not hide their surprise at the mathematical precision of processes and convergences which contributed to the appearance of life on earth and, evidently, in the universe."

Upon watching the movements of a tiny ant under a magnifying glass, I wondered about its joints and struggled to imagine its insides – viscera, brain and metabolism, all so utterly small. My son Gabriel, who was then a university student in Biomedicine, explained quite naturally to me that "we and the ants are made of molecules and it is possible to do anything with a molecule." As an engineer who is accustomed to measuring everything in Nature, I searched in books to find the dimensions of a molecule. My brain and I have lived in a world where things are large. On this subject, we tried to understand some numbers and we sincerely failed, not out of skepticism or unbelief, but out of pure childish naiveté in the face of such greatness of creation. In other words, a single molecule is the size of a single grain of sand divided into 21 billion billion parts.

George Wald, a winner of the 1967 Nobel Prize for Medicine, put it this way, "The most complicated machine created by man is nothing more than a toy when compared to the simplest organism," and he continues, "the greatest difficulty lies in the minute act of a molecule being adjusted so as to become another one, a <u>feat that is not within reach of any chemist</u>."

In 1936, in his book *The Origin of Life* the Russian biochemist Alexander Oparin said that amino acids and other compounds were produced "in the primitive atmosphere of the earth, then composed of ammonia, methane, hydrogen and water vapor." Christian de Duve, author of the book *Vital Dust: Life as a Cosmic Imperative*, argued that the earth's primitive atmosphere was "watery, hot, and acidic." Although in his studies at the University of Chicago the American chemist Stanley Miller was able to produce amino acids in a laboratory, simulating the early atmosphere of the earth, other fellow scientists say that the original atmosphere was not exactly the same as the one described by Oparin.: Still on the subject of these primordial gases, about this atmosphere of the earth's beginnings, there is a relevant fact in another passage from the book of Genesis that describes the following:

"But there went up a mist from the earth, and watered the whole face of the ground." (Genesis 2:6) How could Moses identify the "mist", whether it was water vapor or the other vapors cited by Oparin? The verb "go up" is technically revealing – [it] "**came up** from the earth", that is, it sprouted or sprung up from the bottom.

Take note of what was said by Claude J. Allègre – a professor at the University of Paris, director of the geochemistry department at the Paris Geophysical Institute, member of the National Academy of Sciences (of the USA) and Stephen H. Schneider, professor of biological sciences at Stanford University and co-director of the Center for Environmental Science and Policy and a member of the National Academy of Sciences (of the USA).

"The planet Earth was probably a target of meteorite bombardment. At that time, its surface must have been occupied by volcanic islands and covered by an atmosphere filled with carbon dioxide (CO_2) and dense cloud cover.

Three billion years ago, its surface may have been impregnated with orange smoke of methane which was produced by the first living organisms."*Volcanic islands* means openings in the earth's crust, through which gases and other elements that spread over the surface come up from the center of the Earth.

Chapter 10

The Sixth Day

"**A**nd God said, Let the earth bring forth the living creature after his kind, cattle, and creeping thing, and beast of the earth after his kind: and it was so.." (Genesis 1:24) Among these we can enumerate dinosaurs, whether as small matrices such as crocodiles and lizards or as huge mutations that would not fit into the ark and would not even be brought into the garden to receive a name.

"And God said, Let us make man in our image, after our likeness... And the Lord God formed man [of] the dust of the ground, and breathed into his nostrils the breath of life; and man became a living soul." (Genesis 1:26, 2:7)These biblical texts establish a very clear distinction – animals are living beings and man is living soul. I do not think it is a random, discriminatory classification, much less a skillful play of words, nor a case of careful religious protectionism which clearly removes man from the animal kingdom, where science has now placed him. . Man has a component that differentiates him radically from animals – he has "**the breath**" while animals do not have it, <u>since they were produced from the earth</u>. It is understandable that science does not have a number to quantify the fabulous diversity of living beings that were produced from the earth, but I can clearly extol the individuality of man's creation. The scriptures do not narrate a "**breath**" in the nostrils of every animal that the earth has produced and that still continues to produce until today, but in every human being that is born, whatever the shape or color, language or nation, rich or poor, there is a "breath" that clearly distinguishes him or her from an animal. There is a story of creating a perfect Man, but there is no record, no story whatsoever, that narrates the creation of a perfect animal! I would remove Man from the Animal Kingdom and create the Human Kingdom, in which I would classify the human species with technical and behavioral criteria. Let's make some considerations. The first fact to ponder about this act of creation narrated in Genesis 2:7 is that the creator did not use a "magic wand" nor did he utter some magic words to make man out of nothing at all. The second fact concerns the raw material of this creation, presented as the dust of the earth. It is worth remembering that in the New Testament Jesus Christ also used "the dust of the earth" to heal a man who had been blind since birth, as he was probably born without the eyeball due to a disease called anophthalmia, which had been formed at that time. In this case, the expression "the dust of the earth" is absolutely literal, for Jesus strangely spit saliva onto the ground composed of a dry powder characteristic of the region and made clay or "mud" as it is called literally. In this episode the creative act narrated in Genesis – dust as a replicative vehicle of the characteristics contained in the "breath" and "spit" – was repeated. An adult human contains up to 63% water, a newborn 69% and a fetus 94%. If we mix dust of the earth with some of this liquid, we will easily get clay. Although the term "mud" has become popular, there is no direct quote of it in Genesis.. The first biblical quote of the word clay is "thou hast made me as the clay" (Job 10:9) "As the clay" means giving the desired shape and appearance, just as a sculptor gives his work the shape that he desires by not leaving any accidental change to chance, or as a potter or brick-maker still does today, creating pots and bricks of various technical characteristics. Some 650 years after Genesis was written, the prophet Isaiah expressed it like this, "But now, o Lord thou [art] our father; we [are] the clay, and thou our potter; and we all [are] the work of thy hand." Here the term clay refers only to the moldable characteristic of man, who is in the hands of God, rather than referring to the raw material of creation.

On the purely scientific plane, much more recent ideas highlight the importance of clay in the consolidation of life on Earth. Alexander Graham Cairns-Smith, who was an organic chemist and molecular biologist at the University of Glasgow, said in his book *Seven Clues to the Origin of Life* that **"clay could be the key to the mystery of how simple organic compounds leapt into the condition of self-replicating genetic material**." According to him, *clay would have been the first genetic substance*, which he calls a **crystal-gene**. These crystals, including those of clay, are self-replicating as it is known. Some biologists believe that **clay** was the medium in which RNA molecules (the ribonucleic acid that transcribes and translates genetic information) were formed. In the development of a human embryo, a single fertilized cell becomes a complex organism of 100 trillion cells, with millions of different forms and functions. The human brain is the most complex structure of all creation and it is the most powerful operating mechanism in the universe. Physicists and microbiologists "do not hide their surprise at the **mathematical precision** of the processes and convergences that contributed to the appearance of life on Earth, and in all probability, in the Universe." In the year 2000, a human genome study proved that all human beings had a single origin, a "genetic Adam".

Charles Darwin (1809-1882) developed a monumental scientific work that caused a revolution in Biology, known by the name of Theory of Natural Selection, which was published in his book "On the Origin of Species" in 1859. Darwin considered the changes suffered by all the species, among which he placed Man, to be a result of successive modifications that occurred over thousands or millions of years by the survival of the fittest. Even so, Darwin himself demonstrated his scientific impartiality and honesty with society upon concluding that

"In the nebulous past, there must have been some unique form of life, from which all else was derived." And he himself wondered, "Where did this original species come from?"

In 1882, at the end of his life he proved to be a man of fantastic scientific intuition, as in a letter which was perhaps his last one he wrote, "... the Principle of Continuity makes it probable that henceforth the Principle of Life will be presented as part or consequence of a <u>General Law</u>."

In 1864, five years after Charles Darwin's publication, at the Sorbonne University in Paris and in front of the most brilliant minds of the time, **Pasteur** uttered the following words, **"Can matter organize itself on its own? In other words, can beings come into the world without parents, without ancestors?"** And he went on cynically to say, "Oh! If we could give matter this force that is called life ... what need would there be to resort to the idea of a primordial creation, before whose mystery we must bow down ..." The "appearance" of modern man was so fast and amazing that if we compute the 4.55 billion years of history, Man would have appeared in the last thirty seconds of this story. There is not even an explanation for this surprisingly sudden "appearance" of Man. Evolutionists cannot explain the so-called "Cambrian Explosion" that occurred 600 million years ago. It was a leap in the amount of fossils found that identified unicellular plants and bacteria, transforming the planet which was overflowing with representatives of each species, of plants and animals. It was an abrupt appearance of new life forms. Where would I go wrong by saying that was the fourth and fifth days of creation?! Darwin thus expressed himself in this regard, "The present case must remain without any explanation, and it may be

accepted against the theories written herein. (He was referring to his own theories) In the same period of time, the earth was transformed by the
presence of flowers in an explosion of colors, emerging shrubs and colorful trees. Furthermore, Darwin again manifested his greatness as a scientist, stating with fabulous scientific impartiality, "It is an abominable mystery," for it was completely incompatible with his theory.In the words of Aristotle (384-322 BC) "there is a System of Causes, with Principles that act under the guidance of God. The whole Universe has a single order that must be unveiled by Man."In the face of so many hypotheses, theses, and even mere opinions, there is nothing in scientists to criticize, whether for their mistakes or their correct answers, for their simple opinions or for their formal theses, as they search tirelessly for the answers to our questions. Darwin, for example, consumed literally all his life and died without responding them.As a speculation, which is also science, how would Darwin write his book if he lived in our day and age? Would he rewrite it with the same conclusions, or would his intelligence, combined with the modern resources that are now available, bring us the desired answers? Men like him who are endowed with a brilliant mind should have the opportunity to live 900 years like Methuselah and Adam himself, because only then would their brains adopt every new technological resource and easily answer us. However, it seems that the short human life is a natural hindrance which was carefully created, precisely so that the mysteries of life and creation are not revealed.

Imagine Galileo with the Hubble telescope!
Imagine Pasteur with an atomic microscope!
Imagine Einstein with the particle accelerator - LHC!
Imagine Darwin getting to know about the genome!

Chapter 11
The matrices

"**A**nd out of the ground the LORD God formed every beast of the field, and every fowl of the air..." (Genesis 2:19)

Once again we use a dictionary to find the most appropriate meaning here for the verb "form": to give shape to (something), to conceive, to imagine, to constitute, to compose, to make, to make, to establish, to determine, to establish, to create. It is deduced from the phrase "out of the ground the LORD God formed" that the planet itself already possessed all the elements that were sufficient and necessary to compose the animals and birds (fowl) that were formed on that "day". All living beings that inhabit the earth are composed of CHONPS, i.e., carbon, hydrogen, oxygen, nitrogen, phosphorus and sulfur. In the human body, only 0.001% is composed of other chemical elements. Nature is basically composed of oxygen, silicon, aluminum, sodium, calcium, iron, manganese and potassium. The raw material of which we are informed that man was created was "the dust of the earth". Regarding "every beast of the field, and every fowl of the air", no raw material is mentioned. As the words every beast of the field, and every fowl of the air were emphasized, our logical reasoning leads us to see them as the matrices of all the species that are known today. By means of genetic mutations, adaptive evolutions or by any other instruments of Nature, all extinct animals and birds and those that still exist have been derived from those which were formed on that fifth day. In my judgment, Charles Darwin's observations were not more complete and complex only due to the lack of technical resources to expand his brilliant capacity for observation and deduction. This is how it was expressed by the professor Armenio Uzunian, who has a Master's Degree in Histology and in Medicine and Biological Sciences in São Paulo and Ernesto Birner, professor of Biology with a degree in Biological Sciences from the University of São Paulo. "Recent studies indicate that the animal kingdom is comprised of 30 to 50 million species. For higher plant forms, new calculations point to somewhere around 500,000 estimated species on the planet Earth. These numbers are very surprising and disparate, which shows that scientists have a better idea of the number of stars that exist in galaxies than of the species that inhabit the earth."

In order to study all this fabulous variety methodically, biology has grouped all living beings as follows: Kingdom, Phylum, Class, Order, Family, Genus, and Species. What would have been the interference of the Creator in each species? Would it be correct to interpret that every genetic detail of each kingdom, phylum, class, order, family, genus, and species has been designed in detail by the creator?

"And out of the ground the LORD God formed every beast of the field, and every fowl of the air; and brought [them] unto Adam to see what he would call them: and whatsoever Adam called every living creature, that [was] the name thereof... And Adam gave names to all cattle, and to the fowl of the air, and to every beast of the field..." (Genesis 2:19-20) It is not coherent to interpret that 50 million species were brought into the garden for man to give them a name. By the same token, 50 million species of animals would not fit in Noah's ark or even the largest ship in the world today. The Creator, with his omnipotence which is often mentioned, could easily create each of them. However, this does not seem to be the message of Genesis. It is more coherent and sensible to interpret the message as the creation of some matrices that evolved based on fundamental principles which were defined and established by the creator himself. There are no evolutionary accidents; rather, there are evolutionary principles which are applied spontaneously by nature. It matches or separates, it selects or eliminates, in accordance with these same principles that make something live or make it die. "Evolution works by introducing new options to organisms and then selecting the advantageous ones." As Dr. Katherine Pollard, a Ph.D. and researcher in comparative genomics at the University of California says, "Evolution can clearly take one step forward and two steps back."

Professor Gerald F. Joyce, Dean of the Scripps Research Institute of California and Molecular Chemist put forward his hypothesis in the following way. "... Life as we know it – based on DNA and Enzyme Proteins, with RNA almost always acting only as a messenger of genetic information – evolved from a simpler prebiotic chemical system..."

And Darwin wrote in his book:

"There is grandeur in this view of life, with its several powers, having been originally breathed into a few forms or into one; and that, whilst this planet has gone cycling on according to the fixed law of gravity, from so simple a beginning endless forms most beautiful and most wonderful have been, and are being, evolved."

The posture of science wanting to create life in the laboratory is fabulously interesting. Please read what Professor Joyce said about this subject. "Our goal is to create life in a laboratory, but to achieve it we need to increase the complexity of an organism so that it can begin to produce a new function instead of just perfecting the one for which it is intended."The mystery of life is the point of greatest disagreement among scientists.. There are currently about 3,000 scientists searching diligently regarding topics involved with the genome of living things and life itself. The brilliant and renowned scientist James D. Watson, known world-wide for his studies on the double-helix model for DNA structure and a Nobel Prize winner in Physiology or Medicine of 1962, expressed himself on a spiritual basis for life in the following way. "I'm curious by nature, I like explanations. And if what was desired was an explanation for life, then this theory is related to its molecular basis. I never believed there was a spiritual basis for it. I was fortunate to have been brought up by a father who had no religious beliefs, and I did not experience this kind of obstacle. My mother was nominally Catholic, but nothing more... There is a certain mystique about life, which is very understandable when one is not a scientist, because it is not entirely possible to see how everything could be summed up to molecules and how one could, from them, attain human consciousness and all our complexity ... All we can say is that we do not believe there is any kind of spirit in a bacterium."

Complete neutrality in scientific research is natural and mandatory, and research must be absolutely free of bias, so that with these virtues there is no prejudice in the interpretation of technical facts. Therefore, by these same operational principles, a scientist does not necessarily have to be an atheist, for in so doing, he already characterizes a bias in his interpretations. A scientist's virtue lies in interpreting nature without any emotional interference; he should just read the answers. This is how John S. Mattick wrote about this subject, referring to it as pages hidden in the book of life. "Assumptions can be dangerous, especially in Science. They generally begin as the most plausible or comfortable interpretation of the facts that we know. However, when their truthfulness cannot be tested immediately and their flaws are not obvious, assumptions often turn into articles of faith, and new observations are forced to fit in with them. In the end, if the volume of problematic information becomes unsustainable, orthodoxy has to crumble."

Taking our feet off the ground just a bit and diving quickly into philosophy, how can you accept that you exist only by chance? How can we make our brain consciously absorb that this whole story is incidental and accidental, without a plan and without a purpose? How can we accept that a tear and a smile, hope and frustration, dreams and reality are merely incidental? Seeking out answers is science, but this quest must have reliable direction. Along with the question, "what is life?", science must go along with another question, "why is there life?" This kind of direction will add values that will broaden the focus by giving even more consistency to the searches, no longer just for a dumb, accidental and meaningless life, but for a life with a purpose. Much more personally, I ask whether it would not be more reliable to have faith in a great god, even without knowing or understanding him, by looking at the night sky or simply receiving a child's smile? Man strives to create life in a laboratory by repeating the word, "from the ground" to form birds and animals, and why not say, to form man from the "dust" of the earth (here "dust" means resources available in the earth itself). So why divert the focus from faith by not believing in divine creation; but rather, by believing in the human ability to create the same work? Our logic, then, will lead us to imagine that man will elaborate his own creationist theory and write it in a black book. Then we will return to the question: who will believe in this "new creation"?. Personally, I believe that science will find the mechanisms for creating molecular life in a laboratory. But what will he have created? Bio-androids? Biological robots? And to this will the name of life be given?

From a molecule to a thought, there is a purpose. From a molecule to man, there is a breath. Man is the breath; not the molecule. The breath is what moves the molecule that hosts man. The breath is the image of the creator. I honestly do not think that science will be satiated. It will dare to "give breath". The intention to be greater than the creator is not simply something that exists now. For the same reason, the history of mankind was changed by the very hands of man, who has never been the same since then – from a garden to a cave and from the cave to stupidity.

Chapter 12
After the sixth day

It is undeniable that all the great scientific dissemination has not been enough to convince humanity about the origins of the earth, animals and men, to the point of generating a deep conviction. For just over fifty years now the study of the genome has been a tool for continuing searches for answers that convince people about it. Dr. Katherine Pollard, a Ph.D and postdoctoral researcher in comparative genomics at the University of California, has participated in sequencing the chimpanzee genome. In 2008 she was awarded a Sloan Research Fellowship in molecular biology when she went on to study the evolution of microorganisms living in the human body. Three billion letters make up the human genome. Between chimpanzee and humans there is a difference of 15 million different letters, accounting for only 1% of the genome. She states, "these 15 million letters are what makes us human."

In the May 2009 issue, Ulisses Capozzoli, who is the editor of the Scientific American magazine, says, "This finding shows how chimpanzees and humans may differ profoundly, even though they exhibit almost identical DNA structures." Comparing DNA sequences of the type called Human Acceleration Region 1 (HAR1) of a human, a chimpanzee and a chicken, he came to the following findings. The difference between a chimpanzee and a chicken is only two letters, while the difference between a chimpanzee and a man is eighteen letters.

It is understood that approximately 90% of DNA is what forms the fundamental characteristics of the inhabitants of this planet, which are two eyes, two ears, one mouth, and other common features, patterns in all species of animals or humans.Other smaller percentages make each species different, and an even smaller percentage makes the human brain unique."In other words, it is not necessary to change the genome very much to create new species," says Dr. Pollard, and she continues, "a good part of DNA has purposes that scientists are only beginning to understand." Dr. Pollard says that the brain "is the most mysterious aspect of human biology."

"The [sophisticated] human brain is well known for differing considerably from the chimpanzee's brain in terms of size, organization and complexity, among other peculiarities."

Gil Ast, a professor in the Department of Human Genetics and Molecular Medicine at the Tel Aviv School of Medicine in Israel wrote, "... another puzzle emerged when the mouse genome was published in 2002. It was revealed that the rodent possesses almost the same number of genes as a man, ... and 88% of that is similar."

"As 99% of the genes of both mice and humans are derived from a common ancestor .... therefore, the question becomes: if there are such small differences between the genomes of humans and mice, what makes us so different?" Professor Gil Ast puts it this way, "... the tremendous diversity between humans, mice and perhaps all mammals could have originated from genomes that are so similar." Science now asserts that the genome contains the information needed for creating and maintaining life. W.Wayt Gibbes, contributing editor of Scientific American magazine, wrote "... the genome would represent the book of heredity, the source code of cells, the map of a life. In fact all these metaphors are misleading ...... the genome is not a static text that has been transmitted over generations; but rather, a biochemical machine of astonishing complexity."

Chapter 13
Truth, fantasy, and opinion

Truth is a fact that does not change and cannot be changed, for it is history that has already been written; it is the unchanging past. A lie is distortion of the truth. It generates disorder and conflicts, and for fools it generates an apparent advantage whose expiration date has already passed. Fantasy is a world that does not exist, but it is not a lie. It is simply a well-known dream that delights the human soul. Opinions are volatile words that are mutant rather than solid. They are not true when they come from an ordinary man. However, when an authority expresses them, they come to be considered a passionate, dangerous truth, albeit with little or no foundation.

When a scientist who is regarded as competent states his opinions, a society with no competence to evaluate them accepts them as fantastic truth. If this scientist says that man came from a monkey, society thinks that we really are very similar! If the same scientist says that the universe can exist without God, society thinks that nobody has ever really seen God! SIf he says that the world can exist only if God exists, then society thinks, "I have always had faith in God."

Opinions can be truth, lies or fantasy. Scientifically speaking, society is based on opinions, because its scientific knowledge is naturally limited. Something that is "truth" today in ten years' time may be a lie, or in scientific terms, an unproven theory that has fallen by the wayside. A fantastic, attractive and fanciful theory is still not true. Therefore, from theory to theory the society of the common brains listens, believing and forgetting, believing and forgetting .So is a theory a lie? No, it is merely an opinion that was accepted as truth because it was presented as truth. Gravity is a truth. Light is a truth. Energy is a truth. The air that holds an airplane up is a truth. The water upon which a ship floats and in which a brick sinks is a truth. A scientific truth may shock society, but a theory does not have this right. In 1876Graham Bell shocked society with "a voice that came from beyond." But he had just invented the phone and it was true. In 1859 Charles Darwin came up with a theory that shocked society at the time and continues to shock it until today, even though it is an unproven theory that is not yet a truth and nobody knows when it will be.

Just a hundred years ago the age of the earth was "calculated" at being between 25 and 75 million years. Currently, the "calculations" indicate that the age is from 4.5 to 5 billion years. Ten years ago the world would not have existed without God. Today it is not necessary, in other words, the world might have appeared without him. Where could it be that the truth is: today, ten years ago, a hundred years ago or in the future?

In fact, if man was ready upon being born, then time was noble, because he hosted life. But for evolution, time was just the mere vehicle that carried the genes to be altered by the cold, the heat, the pressure, the sea, and the air.

Science is comfort to mankind. It should not bring about conflict and disorder in society; rather, it should comfort man and dominate this planet. True science dominates the earth - whereas false science is in conflict with man.

Chapter 14
A special perception

The unceasing search for answers to all areas of human existence ought not be confused with vain curiosity. In this pursuit, balance has enabled man to maintain wisdom and lucidity, thus freeing him from becoming his own elementary, uncontrollable hostage. Logic is the tool of balance, and that is what makes mere curiosity different than scientific pursuit. Logic is the compass of science, which holds it firm and stable in every course that is wisely established.

The JWST Project – the James Webb Space Telescope, as successor to the spectacular and successful Hubble Telescope, "...will be a powerful time machine with infrared vision that will peer back over 13.5 billion years to see the first stars and galaxies forming out of the darkness of the early universe".The orbit of this fabulous telescope will be located 1.6 million kilometers from Earth. Just to put this in the proper perspective, the Moon is about 300,000 kilometers from Earth. The instruments of the telescope, which are sensitive to infrared radiation, are able to "perceive the first rays of light", thus being different from the Hubble Telescope, which basically observes light that is visible.

In order to explain how this works let us imagine a showerhead located eighty feet above the floor. Upon opening the valve, the water that comes out of it will take a few seconds to reach the person taking a shower. Likewise, when the valve gets closed, for a few seconds there will still be water falling until it stops completely. Upon looking up at the sky in the nighttime we can see around six thousand stars that are visible to the naked eye. This does not mean that they all still exist. We have only seen the light that came to our eyes. When one of the stars goes out, it will still be visible to us here on earth for many years. I, personally ,consider this project to be some lovely scientific petulance, as it is an apparent delusion of abyssal dimensions; however, it is astonishingly real, doable and fascinating. Man cannot yet see what the universe is showing. As a natural consequence of this fascination, my brain provokes me with the famous, undisputable phrase narrated in the book of Genesis, "Let there be light." Although biblical proof is not the goal of this project, at least it has not been declared as the goal thereof, I believe it will naturally come as a consequence of this fabulous scientific endeavor.

I see that for finding the origin of life, science is on the wrong path, and I justify my point of view with another biblical quote, "Only be sure that thou eat not the blood: for the blood is the life; and thou mayest not eat the life with the flesh" Deuteronomy 12:23. With an atomic microscope one can enter a frighteningly small dimension – one meter divided into one billion parts. Blood is like an ocean in which man is still wading in knee-deep water.

However, the scope resides not in the instruments, but in the ability to interpret what they already show. The secret of science is observation, and it is there, in my view, that all the secrets of life and the universe reside. The perception of a child viewing the facts of life is different from that of a teenager. For a lucid and intelligent elderly person, life is even more different. He sees what younger ones do not see and what he himself did not see when he was fifteen. Although this paragraph has a philosophical tone as we accept the undeniable fact that there are different ways of seeing life, one day a different perception will see the life that blood has exhibited for centuries. Furthermore, even the dazzling universe, whose colors and shapes are available to anyone, will be seen in a different way by someone, who will then see what no one has ever been able to see. What is missing is a perception, just as someone who has this perception is missing. It may be that one has to give up his or her own ideas and become open to new ideas without any meaning. What is missing is a perception like Kepler, Pasteur, and Einstein had, who despite their scarce technical and financial resources looked at life and the universe as nobody else had ever done before. It will not be the instruments that reveal the secrets of life to us. It will be the special perception of an ordinary man. However, to have this perception, a man needs to renounce his scientific beliefs and resign himself to his human limitation. This perception may come from a privileged brain associated with a supernatural revelation. Quite frankly, I recognize that this is a scientific heresy, a technically unacceptable path. Honestly though, even if it was done in hiding, who has never said a prayer? I know not whether Einstein or Pasteur ever did, but Kepler surely did, for he was a declared Christian who, before becoming a scientist, longed to be a pastor. Einstein wanted to discover the formula that God had used to create the universe. I do not know if the god he referred to was the God of the Bible.

At any rate, the supernatural must have been present in his laboratory.

Chapter 15
The sky of the poets

In my endless travels and in most of the nocturnal ones, I always look at the sky in a contemplative attitude. Far away from the city lights, on a narrow, deserted highway amid fields of wheat or soybeans, to the sound of crickets and other nocturnal animals that create the framework of the starry sky, I leave the road, stop my car, turn off the headlights and radio, and then I get out to savor this spectacle, whose price is no more than being bit sensitive regarding the greatness of man and the universe. On warm nights the fireflies seem like shooting stars or perhaps an unknown star that has risen recently, flaunting its light and misleading men who look up at the sky.

Breathing deeply I inhale the cool, fresh air that for a few seconds has dizzied me by its purity or for other reasons which I do not know. I utter not a single word so as not to miss a single note from the musicians who make up the nighttime orchestra; a cricket, a frog, an owl, or the distant barking of a watchful dog. I do not want to defile nature with my voice as though I were not part of it. I let my soul express itself through silent emotions that speak a language which only Nature understands. At these unequalled moments, as the celestial scenery changes every second and its music changes its tones, I become momentarily devoid of mathematics, of physics, and also of business. I forget calculations, profits or losses. By the way, it is more correct to say that the sky calms me down.

Throughout the ages, astronomers have always been in love with it. There is a charming mystery that attracts them passionately every night, casting a spell with the desire to know its charms and mysteries, as though the sky were a mysterious woman who has stunning, untouchable beauty.

The Moon of lovers and poets, whose bright glow has inspired so many dreams and poems, has become an ugly, bumpy ball in the sky. Whereas it was once a symbol of beauty, Venus has been transformed into a dead, charmless planet that exists under a hellish temperature of approximately 467° Celsius. Stars that once adorned the night of men are nothing today but lifeless fireballs with temperatures of millions of degrees Celsius at their core.

Science has taken it upon itself to erase the brightness of the sky and has messed with the feelings of men, discolored dreams, and nullified the imagination. The very science that searches the universe also mocks poets and those who have faith, albeit without being able to add one more day to the life of a dying man.

Fortunately, humanity still has unselfish scientists who are down to Earth and whose ears are within the reach of an ill person's moans and groans. Their eyes remain focused on the reality of men who seek their daily bread, moved by the mockery of their faith, and the hope that is renewed each and every new day.

Knowledge is of no real use unless it is transformed into practical results for men. Knowledge that feeds vanity is selfish and useless. Kepler revealed to the world that which was contrary to his own beliefs.

Chapter 16
The Invisible

What can one say to a scientist to bring him back from forgetting the creator of the universe and on to make him a collaborator in his laboratory? How can one speak of a creator upon seeing the universe and all the life within it as a fabulous accident without one that caused it? So how can one lead him to believe that there is possibly a creator who caused this "accident"?

After all, who has proof that one does not exist? An evolutionist does not see one, nor does a creationist. Therefore, can the argument of such reasoning be "not seeing one"? Nevertheless, we do not see magnetism, we do not see electricity running through wires, we do not see any columns that support an airplane in its flight, we do not see the image of the television that is literally by our side and everywhere, in the sky and on earth, invisible and silent until the moment you turn the TV on. We cannot see the energy that is in gasoline, nor the energy that the wind has within it. Love and hate are also invisible.

Not seeing God is not a scientific argument for denying the existence thereof. Who can see Him? He is only disguised as handsome, ugly, fat, thin, cheerful, sad, rich or poor.

For a creationist the invisible had already existed before the world began. The celebrated phrase, "let there be light" brought the invisible to visibility.

Despite the fact that they are invisible, it is possible to measure electricity, magnetism, energy and use them with full control in favor of human comfort. However, one cannot measure faith. It is not something of physical grandeur. For example, you cannot use it the same way that you use electricity. Faith is not a vector that has module, direction, and meaning. One cannot say that someone has an ounce or a pound of faith. Nevertheless, a person can build a life upon the principles that invisible faith places at the disposal of humanity. It is upon these principles, which in engineering terms are like deep pilings in weak soil, that one builds a life or an existence, with proven, tangible, concrete results.

Faith and Science are two absolutely distinct worlds, albeit with something in common – results. Scientists and men of faith stubbornly seek results as a natural consequence of their thinking.

What can you tell religious leaders to have them get a scientist to leave his oblivion, thus making him a member of their churches or at least someone with whom to share their tiny faith?

They must hear that there is an invisible reason for all things, whether in science or in faith. For example, why is Earth blue when viewed from space? Why is the night sky dark when seen from Earth? Why do things fall? Why does a brick sink while a huge ship floats? Why have the moon and the sun been in their well-known paths near Earth for thousands or millions of years? Why have they not moved? What holds them to Earth? Why does the world turn endlessly? Why do the oceans have their routes? Why is nothing in the sky stationary? Why do we love? Why do we hate? Why do we cry?

Throughout his brief existence, man has had an endless list of questions. Science tenaciously searches for all sorts of answers. However, the man of faith rests passively in the certainty that his God has answers to everything. For him, faith is an accepted truth without material proof. For science, however, proof is the reason for its own existence.

Regarding the mysteries of faith, how is it possible to explain them? How is it possible to explain Mary becoming pregnant without any sperm? How is it possible to explain the parting of the sea? How is it possible to explain the world being entirely flooded? How is it possible to explain the resurrection of a dead man? Do miracles explain everything? Faith wholeheartedly accepts them as miracles which are scientifically inexplicable yet real, but science seeks an explanation of how a "miracle" can physically occur, and there is no heresy in this way of thinking. After all, in a physical sense, everything on earth has a reason. Something that does not yet have an answer has an underlying principle for its way of acting and being which is still unknown, for example, electricity. Mankind uses and controls it without knowing why the electrons move in a given direction. Science just wants to know how nature works, and coherently does not want to give credit to anyone, whether divine or human, without indisputable proof.

Science must follow its path and continue seeking answers to endless questions. Likewise, people of faith must continue to preach that there is an invisible reason, because that provides an answer for everything and for all who believe. Both faith and science must keep to their courses by wisely refraining from any conflict with each other and without any pretension to possessing a truth for which neither of them has irrefutable material evidence.

In this way, faith and science will always be worthy of credit and respect on the part of society. Therefore, we mortals who are inexplicably alive can enjoy what these two powerful tools can wisely give us – physical comfort as well as comfort for our souls.

Chapter 17
Kepler, Newton and Einstein in 2020

Johannes Kepler was an astronomer and mathematician who was born in Germany in 1571. He formulated the three fundamental laws of planetary motion, hence becoming recognized worldwide as the father of astronomy.

Isaac Newton was born in London on January 4, 1643. In 1687 he published his work in which he formulated the law of universal gravitation and the three fundamental laws of classical mechanics.

Albert Einstein was born in Germany on March 14, 1879. He is undoubtedly the most famous scientist, who gained notoriety in 1905 when he published the most fundamental scientific articles of physics, the theory of general relativity, the special theory of relativity, Brownian motion, photoelectric effect, the mass-energy equivalence formula, among other accomplishments. In 1921 he won the Nobel Prize in Physics, and he also received other honors.

Kepler, Newton, and Einstein are three names that have revolutionized the science of their time and of all time. Upon looking at the splendor of a night sky dotted with stars and celestial bodies, taking my calculator in hand for quick and accurate mathematical answers, slinging my computer over my shoulder, or connecting it to the Web for information for my professional demands as an engineer, anywhere on the planet, I have often wondered how they succeeded in their scientific work without the financial and technological resources available to science today. If it were possible to add up what Pasteur, Einstein, Kepler, and Newton spent on their laboratories, we would be shocked by the ridiculously low amounts they had for their research compared to the numbers that are currently made available to science by governments. According to the Bible man was commanded to "subdue the earth." This directly implies that there would be scientific study of every order for the purpose of dominating the earth, discovering all the laws that regulate it and making these laws be in the domain of man for his comfort.

Abusing my imagination, I delight in imagining Kepler with the Hubble radio telescope at his disposal for scientific production today. I close my eyes, take a deep breath, and my mind takes me into a fabulous fantasy – Albert Einstein in an elevator, going down to 175 meters of depth in the Largest Hadron Collider in Switzerland. My mind places Kepler, Einstein, and Newton in our modern days, sitting at the same table and producing science with the fantastic amount of information about the universe that is currently available. Brains like those should be eternal... in a colossal exercise of abstraction, brains like theirs should be transplanted for young scientists to continue their work. None of them committed the foolishness and folly of thinking of themselves as an evolved animal, for by so doing they would be inconsistent with the greatness of their brains, even as we were unaware of the origins of life and the universe. In their life stories, they did not question the existence of a god that was a creator, nor did they express their lack of faith to produce their studies. Kepler, quite the opposite, was openly Protestant. Einstein imagined that God would have used only one formula to create the world.

Common sense is not technical proof. However, the absence of it has caused wars, but its advice has edified humanity.

Chapter 18
The image and likeness

"And God said, Let us make man in our image, after our likenes:..."Gen. 1:26 This sentence is one of the greatest questions of humans who believes in a god as the creator or those who disbelieve. The believer tries to fit in with a god who is the creator. The disbeliever tries to adapt god the creator to be like him. However, both are unsuccessful in their intentions because divinity is not limited to man. Jesus Christ was the most questioned character on this planet, and although he came in flesh and blood and had stomach aches like any mortal, he was not limited by death nor by the greatness of Classical Physics, which was demonstrated by walking on water and performing miracles.

Image, according to a dictionary of the English language, means a copy, something that evokes a particular person or thing, a memory, a simple memory, or even a concrete manifestation of something that is invisible.

Likeness means a thing of the same nature or something similar.Thus, image and likeness have a far broader meaning than a simple and visible similar physical aspect, although this is a common view without, therefore, losing its relevance. Theological and philosophical questions aside, what is the scientific importance of a sentence like this?

In his book "The Descent of Man, and Selection in Relation to Sex", Charles Darwin established a scientific theory whereby "man appeared as a simple animal, fighting and living in a world he had to conquer and was not given as a gift to reign over."

Thomas Henry Huxley wrote his book "Evidence as to Man's place in Nature" in which he expounded on Darwin's theories. In his book "Wonders of Life", Haeckel launched a philosophical theory called Monism in which he claimed that Monera (a colloidal substance) was the "first element from which plants and animals originated." However, he did not explain how this substance that was already alive came into being.

Herbert Spencer, the philosopher of evolutionism, states in his book "The Foundations of Evolutionary Morality" that evolutionism is "the great law of the development of the world and of humanity."

What was the main argument of Haeckel (1843-1919), Spencer, and Thomas H. Huxley (1825–1895) for claiming that humans had descended from primate?

It was recognizably the likeness.

They did not have the powerful technical resources that are currently available to the most renowned researchers. As they were limited in such a way, human grandeur was unfortunately included among the existing animal species and compared by sheer image and resemblance without the requisite scientific rigor. In other words, by image and likeness man was compared to a monkey.

Why is physical likeness to a monkey given social and scientific credit while divine likeness is discredited?

The answer seems to be elementary, in that a monkey is visible and God is invisible. In my point of view, the true answer does not go through the path of divine invisibility. Throughout its history, humanity has created images of hundreds of gods that had a human aspect. Why are gods similar to man? Whether consciously or unconsciously, man is projecting himself into this image without any spot or blemish, without sin, a man full in his actions and emotions, that is, God.

This projected image of a state of fullness externalizes the conviction of a currently decadent situation. In other words, it is the recognition of an altered state of man in relation to the original one that he once had. Children are similar to their parents. Friends are not similar to one other physically, but in their attitudes and goals they are.

The words image and likeness have accompanied man ever since day one, which is the day that never existed according to evolutionists. For them, nothingness was transformed into something, which without explanation turned into life, which without explanation turned into a monkey, and which also without explanation or proof perhaps became man. In short, for evolutionists, a smile, love, solidarity, affection, kindness, hatred and evil arose spontaneously, without being caused by anything, without any purpose, plan or principle, just incidentally and accidentally out of nothing Therefore similarity does not exist either.. Man's projection into a god is foolish and without origin. Sin and wholesomeness do not exist either. A form of man that was original, wholesome and wonderful did not exist either. Nor did awareness of sin, of mistakes, of the need for correcting ones course during their lifetime ever exist either.

For evolutionists, life is nothing more than a purposeless accident, and men are nothing more than more developed animals whose lives, as a result, have no purpose either.

For them, we humans are in a jail called Earth which is only feeding us, allowing us to reproduce, and by this logic, evolving stupidly and unconsciously until death.

For Creationists, however, life is an opportunity given to man to experience divinity, to love, to be happy and to create. Life on this enchanting planet, in this garden called Earth, is an opportunity to be like God which is given only to man.

Chapter 19
The scientist

The secret of science is observation and logic is its compass. Emotions or even discrete bias can invisibly disguise the interpretation of a scientific observation. How many apples fell without anyone questioning the phenomenon? Nonetheless, when one fell on Isaac Newton's head he wondered how and why it fell. Atheism is a form of religious bias just as genuine faith is. When faced with such a categorical statement as "Let there be light" a believer would exclaim, "how powerful God is!" An unbeliever would firmly state, "that is impossible." A scientist would ask, "how could this be possible?" He would neither doubt it nor believe it. His logic would lead him to the impartiality that neither the atheist nor the believer possess. As a matter of principle he doubts nothing but he questions everything. In his research, success depends on his ability to observe, not on his faith nor his disbelief. And that is what science is all about: proof and impartiality, even though proof will damage one's faith as was the case with Kepler, who proved that the earth was not the center of the universe. Yet his faith was not shaken, for he was reportedly a Christian scientist. And what about Charles Darwin, who had studied theology at Cambridge?! The difference is the interpretation of what one sees, for Nature is sovereign and it always has the last word, wholly indifferent to faith or unbelief. Nature simply is.

Chapter 20
Adam's brain

Why do we use only 20% of our brain? When or where will we use 100% of its fabulous capacity? Could our brain be smaller? Does our brief existence on earth not demand its great size?

We can interpret these questions with the following sentence. Our powerful brain was not made for such a mediocre life as chasing down our daily bread. Strange as it may seem, this is a conclusion to be coldly considered.

By definition, evolution is development, it is the enrichment and improvement of a species with successive progress towards the full adaptation of a living being to an environment or situation of life. From this point of view, the monkey was already fully adapted to its habitat, which is its forest. Its four legs, which function as arms and hands, and its tail had given it perfect mobility and safety in the treetops. Why then would the monkey evolve in physical appearance and keep its brain with the same capacity that is limited to the daily search for food and shelter? A fish with a tiny brain does the same thing with a two-second memory. What evolution was it that has limited man's brain to 20% of its full potential?

According to the bible, Adam spoke to God and saw him daily "in the cool of the day", which I suppose means late in the afternoon. This certainly required him to have a complete brain; after all he was "the image and likeness" of the God of the bible. Talking with one who has a brilliant mind requires much greater intellectual capacity than what is usually needed. It is easy to understand that a common worker in his routine tasks demands minimal usage of his brain, which is quite different from the everyday life of someone who has a complex activity that demands a brain that is already insufficient when limited to its 20% that is available or usable .

If we disregard the biblical Adam with 100% of his brain, an indispensable condition for daily contact with eternity and perfection, and turn to the monkey, we can conclude that the brain has kept its capacity. Since the human brain had its dimensions increased when compared to that of the monkey, we can easily deduce that evolution despised 80% of this organ.

Regarding this subject I know no answer from Darwin or Lamarck, for Darwin himself said that in order for his theory to proceed, an "original species" would have been necessary to start and evolve from it.

Chapter 21
Has evolution stopped?

Everything we know has a beginning, a middle and an end. The bible itself begins with "In the beginning ..." and it consistently describes an Apocalypse. The great empires did not escape this rule, even though no one had written it. Man himself knows his beginning, lives in his environment, and knows how his story will end.

The history of the universe, however, is literally crammed with unknowns that have solemnly roamed through every age, through every lab, and through every temple ever since the foundation of the world, arrogantly sweeping the present and heading gaudily and imposingly into the future without anyone being able to slow them down with answers. Has evolution stopped with the mankind and life of the 21st century? Just to exercise our speculative capacity, if the answer is yes, another question that will immediately burst forth is, "who or what gave the signal for it to stop?" If the answer is no, some fascinating questions that will silence mankind are "what will we men be transformed into in the future, and what will the great invisible force called Evolution do to us?" This goes without questioning the evolutionary destiny of all other life forms, of which there are around forty million.

According to this force, something that was primary was completely ignored until it turned into something secondary, then tertiary, and so on until the monkey appeared, and the monkey became man. Therefore, because we are an evolutionary transformation of what had already existed, according to this reasoning, in a future with no expiration date, can a fish become a man with scales and gills? Can an eagle or condor become a winged man who will fly just as easily as they do? Those questions seem silly, like fantasies. But who can dispute them? If the unseen force of nothingness made a monkey evolve into a man, who can deny that this same force may act upon a fish or a bird to produce a flying man or a man from the depths of the sea with the same vigor and voracity?

If evolution is an endless process, what are we men transforming (or using the same verb, evolving) into? What will we be like in the upcoming "millions" of years? Will we remain human or will the transformation of climate on Earth make us grow hair like gorillas or polar bears? Will hatred as well as social and commercial competition bring us back to the caves? Will technological evolution increase the dimensions of our brains and our skulls as well? Will our hands have longer fingers as we routinely use our computers? Will continual wars cause horns of evil to grow on our heads? Will Earth's soil, pockmarked by bombs, create evolved bipartite hooves from our beautiful feet that balance and support us?

Let the evolutionists answer the questions above.

Please don't take me for a fool after having read these lines. I am just striving to follow the same logic as the evolutionists, who give credit to all kinds of deductions and speculations that come from a fossil. One archaeologist has stated that all the fossils that have been found prove only 10% of the history of mankind, and this 10% provides the basis of all possible and imaginable theories, theses and hypotheses regarding the history of mankind and the universe.

The sea does not stop, for all day long and night long its waves break on the beach. Why? Is it another evolutionary accident! After all, is everything therefore an accident? This is a comfortable so-called scientific explanation, albeit without a mathematical equation, so how can it be proven? Neither creationists nor evolutionists *are able to, so both go unanswered.*

Both settle for what they believe. Neither can accuse, for after all, who has the proof?

One need not be an atheist to be a scientist. To be an atheist is to be partial, to be biased, because it takes from scientific research a path that will not even be investigated, which is motivated by prejudice and discrimination – "a medieval form of obscurantism", in the literal expression of an atheist. It is a wisdom-like bias that pushes a scientist off balance. Einstein had a formula for invisible energy, and Isaac Newton had a formula for the invisible forces of mechanics. Kepler had a formula for the likewise invisible and fabulous forces that move celestial bodies.

Chapter 22
Natural miracles

A dictionary of the English language defines a miracle as a feat or an extraordinary occurrence that is not explained by the laws of nature. For example, we can cite cures for diseases, for the sea opening for a people to pass through it, for walls crumbling down, or for a rain of fire falling on a city, among so many other things. The bible is full of narratives of such events.

Just for the sake of reasoning, allow us to accept the existence of a god as the creator, albeit without the classical questions. In this way we will find easy answers to some of these miracles, for this god the creator is omnipotent, omnipresent, and omniscient, thus having dominion over all the forces of nature. In other words, damming rivers or causing earthquakes at specified times and places is not a problem. For the creator, the condition for these and other phenomena to happen has always been only one – obedience to him. Science has already found explanations for many of these miracles and I particularly do not doubt it. However, science does not recognize them as miracles, as natural phenomena explain how the waters of the Red Sea were held back. Not recognizing this episode as a direct divine interference, it will remain for science and those who think in a like manner to qualify it as an act of luck for a large number of people to be at that very moment on the beach at the very moment that the sea casually opened before them, for enough time yet for them to casually make their crossing. Likewise, another stroke of luck and of the same magnitude was the closing of the same sea right upon the enemies that were pursuing them in that same scenario.

Following the same logic, an earthquake which struck the city of Jericho destroyed its protective walls, and that also occurred just as a people passed by wishing to enter the fortified city without having the weapons capable of such a war operation.

The cities of Sodom and Gomorrah rained fire and brimstone from heaven. In the surrounding soil there were, according to studies, deposits of methane gas, tar and other elements capable of producing fabulous explosions.

With abundant supporting material, deserts and high mountains are flooded with fossils of marine animals. For its part, science is silent on the primordial oceans that completely flooded the earth in its early days and the controversial biblical flood.

There are impressive studies that describe ways or means by which these "miracles" may have occurred. In other words, science is finding the tools used by the god of the bible. The historical fact is, in my opinion, more important than the tool. Whether it was a miracle or a phenomenon with natural causes, the fact occurred and it is part of history.

The question then ceases to be whether the fact occurred and becomes who caused it, being open to anyone who wants evidence and more evident to those who have faith.

The richness of faith is believing without seeing, it is waiting and receiving, while the richness of science is believing what is proved and enjoying what has been proven. The honor of the scientist is the proof he sees. The believer's honor is the faith in that which he does not see. They are two distinct worlds with noble goals. Faith is personal and science is collective. Faith is for the soul and science is for the body. They are two indispensable worlds with a long history of difficult coexistence. These are two proud worlds that could be just one, therefore stronger and more useful.

Chapter 23
Involution

The English language dictionary defines "involution" as the opposite of "evolution". In the academic and scientific world the word evolution is easily associated with Charles Darwin, but other scientists in their studies of the origins of the universe and humanity have gone far beyond it. His theory, according to his own words, is based on there being a primordial being of each species, from which to evolve.Making an involution on the basis of classical theories and the latest theories, starting from modern man and passing through the monkey, we will come to nothing. This reasoning is profoundly degrading because it leads humanity to believe that once human grandeur was a wild animal and before that it was nothingness. According to science, the universe and man are but an accident, a meaningless and purposeless accident. Given this theory I firmly believe that even the most atheistic man would question himself saying: "God does not exist, but I exist and I am real." He would confront nothingness with man, nothingness with life, and nothingness with love. For him it would no longer be a matter of faith, but of a simple logical view of facts and values that are lived daily by every human being. He would still be satisfied with the scientific explanation for his atheism, but not conforming to it, he would go on rambling: I was nothing, then I was an animal, today I am nothing, and I will keep being nothing, without any reason for my existence!

Chapter 24
"Tinker Bell"

It is not necessary to be a calculating engineer or a geologist to conclude that an earthquake requires a great amount of force to cause it. An elementary practical experiment can easily prove this statement. When we punch a table, all objects on it will move in proportion to their proximity to the point of impact. According to Newton's Third Law, the force applied to the table will be equal to what the table will apply to all objects on it. An earthquake is caused by an underground displacement of rocks which, upon moving, cause shock waves to reach the surface of the earth.

It has been argued to me more than once that a miracle can occur without a causative agent and without a physical cause. Putting our imagination to work, I wonder how the Tinker Bell would trigger an earthquake with her magic wand? Abusing our fantasy, let us imagine her near the walls of the city of Jericho with a job to do: to knock down those walls when the people who are going around it complete the seventh lap. The exact moment will be when that multitude of people shout in unison. Tinker Bell was alert, carefully counting each completed lap. Upon lap six she tried to move a small part of that wall to test her wand, but nothing happened, nothing moved or gave any sign of doing so. She then reasoned, "not a stone has left the place, for these walls are very strong. Something will have to get these stones out of balance." Then she had an idea. Only an enormous amount of force will be able to move the wall. The seventh lap was almost over and she was not yet sure what to do. She was flying hurriedly around the wall, when suddenly she heard a shout. "Right now!", she exclaimed nervously. Instantly she plunged into the ground, went deep, waved her wand and spread her magic dust vigorously.

Instantly she heard something like thunder, and everything moved. Upon returning to the surface, she saw the walls on the ground and the people entering the city over the fallen wall. Walls do not collapse without a cause. Rivers and seas are not dammed without a cause. The waters of a moving river have kinetic energy, that is, the energy of motion. It is this energy that drives the turbines of a hydroelectric power plant which transforms it into mechanical energy, which being connected to a generator is transformed into electrical energy for our homes. To stop a river, a small stream or an ocean current, a force equal to the movement of its waters is indispensable. Thus the Red Sea or the Jordan River was interrupted in its respective course. Was that a miracle? Yes, it was a miracle performed with the forces of nature itself, with the forces of creation itself. But what about curing a disease in the human body? The reasoning is the same. An antibiotic is a biological tool used by Medicine when it knows the causative agent of the disease and how to overcome it. Like an earthquake, it is the use of a force that creation itself produced. The hemorrhaging woman, the cripple at the temple door who Peter raised, the blind Bartimaeus, and so many other miracles described in the bible are all very clear testimonies of nature's actions in favor of nature that are performed by those who know this nature. When medicine knows the human body in depth like the one that created it, all diseases will be curable, for new biological weapons will combat the evils of the body. As long as this does not happen, believers have the best and most efficient natural weapon – faith in the one who knows what man does not yet know.

Science is analyzing the force used, for how would the earth tremble without a powerful force at work? The issue is no longer a scientific questioning of faith, and it coldly becomes no longer the cause but that which causes it. Whether it was a fabulous coincidence of time and place for the people to be passing with that intention, or whether it was Tinker Bell with her magic wand, or it was the god of the bible, the issue is only who or what caused it to occur.   But then, who did cause it? To whom does faith attribute it? To the God of the bible. To whom does science attribute it? Science doesn't even venture a guess. To whom does fantasy attribute it? To Tinker Bell and her magic wand.

Chapter 25
The first man

In scientific terms, this man is not yet known. His age and appearance, his habits and his intelligence are mysteries that have always stood the test of time without any answer or clue. The Neanderthal Man, who inhabited western Asia and Europe about 300,000 years ago, according to studies of his Mitochondrial DNA, does not belong to the human lineage. However, recent studies have indicated the presence in them of the same gene responsible for speech that we have. Regarding an artistic expression found in a cave, there are opinions that qualify them as beings who already expressed themselves through art. They are mysteries that add to the many existing ones that continue to challenge some selfless scientists. The famous "missing link" that would make the connection between the monkey's evolutionary chain and man has not yet been found.

Upon leaving the home of his parents, Adam and Eve, Cain began to wander the earth, for he had been born outside the garden. According to this biblical narrative, at this time in human history "there were giants in in the earth in those days."

But when science speaks of the first man, what does it look for or what does it expect to find? Without being poetic or philosophical, what would characterize this "Adam"? Would it be his physical appearance, the evidence of his family life, social organization, his clothing, or perhaps some artistic manifestations? Undoubtedly, any of these items, even separately, would be a consistent indicator. But man is the sum of these elements and so many others. Therefore, science cannot expect anything different from what we are today, technological issues notwithstanding, of course.

From the perspective of evolution, it is unknown what one would look for. Everything becomes a clue: a longer femur, a different joint, perhaps a larger or smaller skull. Albeit with some bias, everything becomes yet more "proof" of evolution, which is the great puzzle that science has insisted on assembling for 150 years, although there are still millions of missing pieces.

Trying to be as rational and logical as possible, freeing myself from emotions and any connection with faith, I see in man the reflection of a deity, in the greatness of emotional extremes ranging from love to hate, from tears to laughter, from hope to discouragement, from victory to failure, and especially in what we call conscience. I do not consider an accidental development of love and hate to be technically possible. Variations in pressure, temperature, climate, humidity, air components, and other elements of nature do not act on the invisible so as to produce that which is invisible. Behind a tear there is pain; behind a smile there is a cheerful heart, and behind a brave initiative there is a dream.

Behind all human existence there is something invisible, from the mystery of pregnancy to the mystery of death. In pregnancy there is a seemingly ordinary world that is forming, and in death a disparate world disappears. Man is not a mere species, man is an individual.

Chapter 26
Adam's age

If we arbitrarily consider the Adam of creationist theory to be the first man on the planet Earth and perhaps in the universe, we will need some elements that are not yet available in order to discover his age and history in relation to us. Regarding his creation, the Bible, the Koran and the Jewish traditions are unanimous, differing only in some aspects of his raw material. None of them, however, tell us the exact date of his "birth". They tell only about a period of history, locating it in the creation timeline, and qualifying him as the first inhabitant, end user, and administrator of the planet. It is known from the sacred texts that he "was born" on the sixth day and spent some time alone, then along came Eve, and then there was still a whole seventh day, whether it was a "day" of 24 hours, a 1000 years or a million years. Before living in Eden they inhabited the earth in conditions that were fully favorable to life and mankind. It was a period of life of unknown duration, of which the bible does not give exact information. After a while, they were invited to live in the garden, where they also spent a period of time until their expulsion. Upon the involuntary departure of both of them, the earth was cursed and at that very moment there was a radical transformation of the planet. The scenario was different, no longer as it was the original project for the existence of all life, especially the human one. As for this transformation of the planet, the only biblical quotation is that the earth would produce "thorns also and thistles", i.e., plagues, difficulties, pains, dislikes, hurts and mortifications, according to the dictionary of the Portuguese language. Consequently, all life on earth was significantly affected. For the first humans, for all the animal kingdom and the vegetable kingdom everything, absolutely everything must have changed in their lives and in their bodies. The bible does not detail Adam's quality of living outside the garden. The fossils or tombs of Adam and Eve have not yet been found.

In the context of science we have no more information than we have in the holy scriptures. On the contrary, the confusion is greatly increased by the fact that evolution adds one more mystery – the monkey. There is no book that narrates the creation of this animal, from whence it came, from whom or from what, nor is there any connection made with its so-called successor, which would be man.

Thus science and religion do not precisely answer this question, which is the greatest one of humanity. Both continue to demand our faith – whether in God or in Science itself. Scientific logic remains insufficient to present a theory with a minimum of sound basis and reliability. Thus, logic remains aside, waiting for its moment to be used, relinquishing its place although it resists faith in the invisible.

Chapter 27
The first caveman

The Adam of the Bible was created to live in a habitat that contained all the ideal resources for his literally perfect existence. The weather was ideal for his physical condition. Its temperature and humidity were perfectly synchronous with the senses of man. Food and crystal clear water were readily available. The earth was made for man.

Upon being expelled from the garden Adam was prepared for absolutely nothing. He could not hunt and could not protect himself from the cold or the heat, because he had not known of them. He could not sew his clothing, for the creator himself had made him the first of them out of animal skins. His body came to suffer the consequences of the new living environment, which was unsuitable for his metabolism and comfort. His teeth began to suffer from the action of new foods and his skin had to adapt to low temperatures and manual labor. His hair grew inordinately and became a form of protection for his head, a way to keep it warm. His hands and feet became coarse from working the fields to prepare the land and plant his food. Uncountable evolutionary changes took place in his body to protect him from the aggressions that the new environment brought about on a daily basis. He had no house, no tools, and he could not build anything because he had never needed to. The weather outside the garden had strong variations in temperature, and the rain and the cold came along. In desperation he migrated at random to escape those climatic conditions. One day, in the midst of a storm, fleeing lightning and falling trees, he accidentally found a cave and took shelter there. But then came snow, something which he had never seen nor even imagined existed, and with it came freezing cold temperatures. The simple shelter of the cave was no longer enough to keep him warm. Freezing, he instinctively began to move. Freezing, he instinctively began to move. Even colder, he ran back and forth within his "home". There was a bunch of dry straw on the floor that he had carried in to form a mattress. In the midst of this disorderly scramble for comfort, he tripped over some whitish stones, causing a slight collapse. The stones sparked as they crashed into each other, igniting the dry straw where they lay. The fire spread without being fought, for he had not known of fire. The flames and heat were stunningly surprising to him, but at least they warmed him up. Months went by and the snow melted. When the heat came he had a different physical appearance. His hair had grown without any care, his hands and feet were thick, and his nails were dirty and long. His skin had become dry and the fine habits he had had in the garden turned into crude ways of mere crude survival. Adam adapted to the world outside the garden. Over time, profound adaptive mutations occurred naturally. Thus, the exquisite first man who had lived in the Garden was transformed into a crude caveman.

The man had been designed and set up to live in a specific "garden". This would have been his natural habitat, with all the right conditions – climate and resources, all of which were available for his comfortable and balanced life on earth. Upon leaving the "garden", his whole configuration collapsed, and from then on, up until this very day, human conflicts multiplied exponentially, reaching levels , of complex and almost unsustainable control. Far from the "garden", along the timeline, living or at least surviving outside of his habitat, his consequent social, physical and intellectual transformation thus became understandably justified.But something remained intact – his magnificent brain which helped him find the most varied solutions for his basic needs, with sticks or stones, with or without fire, inside a cave or in a comfortable air-conditioned house with.

frozen food. That is what we know, or at least it is what we think we know!

Chapter 28
Evolutionary Creation

Evolutionary Theory presented the tools that Nature used in primordial Creation. Nature is the sum of all the principles governing the functioning of the universe. These are the fixed and unchanging laws that science knows acting simultaneously with those which are yet unknown.

Faith presents answers with few letters, even because it does not require material proof, leading man to believe in an intangible and historical testimony.

Science, in turn, is made with a good hypothesis, its respective research and the consequent evaluation of the results. It is absolutely tangible in some ways and speculative for its sometimes miraculous deductions and cinematic creativity.

The coexistence between them has been turbulent, motivated mainly by the absolute lack of reciprocal technical knowledge, a scientist who does not know the scriptures, and a theologian who does not know science; a scientist who claims to be impartial in his studies and a theologian who is embarrassed by his skepticism regarding science.

Adam, according to the scriptures, was the first man. His creator had configured him for a specific purpose, just as an architect or an engineer creates his work for a well-defined purpose. If Adam's creation is fabulously incomprehensible to mankind, this creation was unique because Eve was created from life which had already existed. So Adam was created in one way, Eve in another, and her descendants, humanity, was created in yet another way. It is then imagined that the secret of human life after Adam lies in man's "rib", for something has been extracted from it for the creation of another life similar to his own. To this end, he was anesthetized, for God "caused a deep sleep to fall upon Adam."Gen.2:21

Chapter 29
Epilogue

Time is the written record of history, in which the phrases of men are written. Some are written in pencil so that time itself erases them, while others have been carefully written in indelible ink, which to the delight of men cannot be erased.

Wisdom and knowledge do not always reside in the same man. Wisdom has answers that knowledge does not know. A wise man's eyes are his brain and his soul. He who has knowledge believes only in what he sees without distrusting his own eyes.

To a man who has had his life rebuilt by his faith, what does it matter whether or not the earth is 4.7 billion years old?

To those who by faith have been healed of a deadly disease, what does it matter if the magnetic north of the earth is not true north?

To someone who, through fantastic high-tech surgical intervention, gained a transplanted organ, does it matter whether a budding star needs two million years to appear in the sky as we see it from earth?

It does matter what science and faith provide mankind: faster planes, more economical cars, new vaccines, a restored family, the opportunity for a fresh start. Hope does matter a lot, whether one hopes there is an eternal god or a mortal scientist. In human history, science has literally moved visible mountains while faith has moved invisible mountains. Both have brought the reality of these two worlds to mankind.

Thus, completely indifferent to historical and hysterical conflicts, modern man sees faith and science as the two pillars of his own existence. For him, the truth lies solely in the practical results of what he believes.

So it is enough for science to develop formulas for it, and it is enough for the church to preach faith Both of them are for man, the magnificent masterpiece of the universe.

Marcelo Caldana is an engineer graduated from the University of the Rio dos Sinos Valley - UNISINOS, Brazil, Rio Grande do Sul, in 1981. Since then he has developed over 400 projects of machines and equipment for the industry. Conducting lectures at universities, high schools, churches and at national and international fairs. He received an industrial award in 1993 and is currently a technical consultant working throughout Brazil on Environmental Technology.

He is the author of the book *"The boy and the rat - an encounter with wisdom"* by PUCRS Publishing House and six other books. Born in Porto Alegre, RS Brazil in 1952, where he lives with his wife Susi Dickin Caldana.

Contact: ecoengenharia@brturbo.com.br

BIBLIOGRAPHY

SITE INOVAÇÃO TECNOLÓGICA. Sonda espacial Corot parte amanhã em busca de vida no espaço. 26/12/2006. Online. Disponível em www.inovacaotecnologica.com.br/noticias/noticia.php?artigo=01 0130061226. Capturado em 21/11/2009.

SCIENTIFIC AMERICAN BRASIL Especial 20 : As formas mutantes da terra, issn 1807-1562

http://www.oceanario.pt/cms/13/

WIKIPEDIA   -
Source=http://ga.water.usgs.gov/edu/watercycle.html |Date= |Author=John M. Evans/USGS-USA Gov |Permission=Free for all use |other_versions= }} Category:Water cycle {{PD-USGov-Interior-USGS})

http://geology.csupomona.edu/drjessey/

({{PD-USGov-Interior-USGS}} courtesy of the U.S. Geological Survey imported from en.wiki)

http://www.ufmt.br/

O UNIVERSO – Isaac Asimov - bloch editores 1969

Sir Roger fez grandes contribuições à relatividade geral e cosmologia, principalmente por seu trabalho sobre buracos negros e o Big Bang"[2].

BIOLOGIA volume único 2ª edição ISBN 85-294-0269-3 Armênio Uzunian e Ernesto Birner

Gil Ast, em Nature Reviews Genetics, vol.5, pág.773-782, Outubro de 2004

https://jwst.nasa.gov/content/science/firstLight.html

Challenging the dogma: the hidden layer of non-protein-coding RNAs in complex organisms. John S.Mattick, em BioEssays, vol.25,nº10, pag.930-939, outubro de 2003.

Edge.org *To arrive at the edge of the world's knowledge, seek out the most complex and sophisticated minds, put them in a room together, and have them ask each other the questions they are asking themselves.*
Sat, May 25, 2019 - Martin Rees article
http://www.vliz.be/docs/Zeecijfers/Origin_of_Species.pdf